New Superconductors

From Granular to High T_c

New Superconductors

From Granular to High T_c

Guy Deutscher
Tel Aviv University, Israel

NEW JERSEY • LONDON • SINGAPORE • BEIJING • SHANGHAI • HONG KONG • TAIPEI • CHENNAI

Published by

World Scientific Publishing Co. Pte. Ltd.
5 Toh Tuck Link, Singapore 596224
USA office: 27 Warren Street, Suite 401-402, Hackensack, NJ 07601
UK office: 57 Shelton Street, Covent Garden, London WC2H 9HE

British Library Cataloguing-in-Publication Data
A catalogue record for this book is available from the British Library.

First published 2006
Reprinted 2007

NEW SUPERCONDUCTORS: FROM GRANULAR TO HIGH T_c

ISBN-13 978-981-02-3089-0
ISBN-10 981-02-3089-3

Printed in Singapore

To Aline, with love

Foreword

Few fields of solid state physics have produced such a worldwide flurry of studies and publications as superconductivity, especially after the two major advances that were the theoretical model of Bardeen, Cooper and Shrieffer (BCS) and the discovery by Bednorz and Muller of the cuprates as high temperature superconductors (HTS). In more recent years, the field has been extending to new materials, notably organic as well as rare earths or uranium compounds; a better knowledge has been gained of the structure of materials and their physical properties, while the extreme variety of proposed models besides BCS's has now converged towards very few types of solutions for HTS materials. But so far, no new type of material has been developed that could be used at room temperature. The industrial production of NbTi, Nb_3Sn or cuprate wires able to sustain large magnetic fields at liquid nitrogen or lower temperatures has made possible the use of magnets for MRI studies as well as for large accelerators. This is presently the main industrial application of moderate or "high" temperature superconductivity; but the large critical currents that can be obtained at low temperatures open the possibility of developing high intensity electric transport in towns or large energy stocking centers.

Guy Deutscher's present book does not attempt to cover the whole of this proteiform field. Being one of the first students of P. G. de Gennes at Orsay and now a Professor at Tel Aviv University, after an initial training as an engineer, he is well versed in both the theoretical and practical aspects of HTS, where he has made many personal contributions of note.

His interest focuses mostly on the cuprates. He explains how they differ from classical low temperature superconductors. He strongly points out that to reach a high critical temperature T_c and a large gap Δ, such materials have very short coherence lengths, down to atomic scale already below room temperature. This, in turn, develops the possible role of thermal fluctuations of superconductivity which are indeed negligible in low temperature superconductors but visible in cuprates, both above and below T_c, for instance, in the heat capacity.

If strong enough, these superconductive fluctuations could transform continuously a strict BCS superconductor, described in a fluctuationless Landau approximation into a more localized Bose Einstein (BE) condensation of electron pairs, where the temperature of pair creation could appear well above the critical temperature of correlation into a state akin to a superfluid such as He. With this idea in mind, Guy Deutscher spends his first chapter comparing BE and BCS condensations; he also describes in detail a similar progressive change he observed in "granular superconductors", where grains of diminishing size of a metal such as Al are exchanging electrons through insulating junctions such as Alumina or Germanium.

These models allow Guy Deutscher to present the current explanation of the "underdoped" region of cuprates as a range where one possibly goes progressively from a BCS superconductor at overdoped and optimally doped samples to a BE model in the underdoped regime, where T_c gets progressively depressed while a "pseudogap" appears at a progressively higher temperature, possibly that of pair formation.

In the same spirit, Guy Deutscher is, I believe, one of the first to suggest that thermal fluctuations can be strongly detrimental to the use of superconducting magnets, possibly well below T_c and, at low temperatures well below H_{c2}, the critical magnetic field for stability of superconductivity. The main reason is that, because of their small lateral dimension, of order of the coherence length, the vortex lines by which the magnetic field penetrates into the superconductor can be fairly easily freed, in their thermal vibrations, from the lattice defects introduced to block their motion; when freed, the vortex lines produce in their motion a strong electrical resistivity. This effect is more developed in bismuthates, of high T_c but small interplanar coherence length, than in YBaCuO, of lower T_c but larger coherence length, hence a stronger pinning of the vortex lines.

Although Guy Deutscher insists on the possible connection of fluctuations in BCS with connected fields of granular superconductivity and BE condensation in superfluids, he is very careful to examine experimental facts in a balanced way and to describe the possible models. He points out that a BCS type of approach seems to work not only for low temperature superconductors but also for the cuprates near the optimum doping. There is in fact, as he says, no special difficulty in explaining the large values observed for T_c and Δ in most of the cases he discusses, if only one uses, as one should, the density of electronic states at Fermi level $N(E_F)$ and not try and express it in terms of the Fermi wave number of the free electron gas. Thus it is well known that the large T_c observed in Nb or NbTi corresponds to the bonding peak of the d band at the Fermi level. More generally, large values of T_c can be related to situations where the Fermi level sits at or very near a van Hove anomaly of the density of states $N(E)$, which comes from a specific diffraction effect of the structure on electrons. Thus in the small grains of granular superconductors, the nearly spherical form of each grain can give rise to electronic states near to highly degenerate spherical harmonics, possibly leading to large values of $N(E_F)$; this might explain the increased value of T_c in the very small grains limit, reached just before the Coulomb blocking of current related to the finite energy distribution of electronic states near the Fermi level. But more important, the author recalls that fairly high T_c's in Nb_3Sn might be connected with a van Hove singularity at the edge of one dimension d bands, while the exceptionally high T_c's observed in cuprates have been related to two dimensions van Hove peak in the middle of the valence band of CuO planes. The Fermi structure in such an approach has indeed been fitted by S. Barisic with band computations by assuming a significant direct electron transfer between neighboring oxygen ions.

As stated in the book, even stronger van Hove anomalies can be obtained in the cuprates if one takes into account the local antiferromagnetism observed to at least optimum doping. The presence of magnetic moments on Cu ions with a local antiferromagnetic order should split the van Hove peak into two stronger one dimension anomalies, somewhat broadened by the antiferromagnetic long range disorder. If one accepts this admittedly qualitative point of view, most of the variation of T_c and Δ can be described with this picture, without the necessity of invoking a tendency to BE coupling. Indeed the peak in $N(E_F)$ predicted at optimum doping was observed early as a Knight shift in NMR studies. Also the pseudogap can be seen as due to one hole excitation to the peak on $N(E)$. If, as NMR studies suggest, local antiferromagnetism persists in the overdoped range, one should observe a certain symmetry between over and underdoping, as indeed is observed in Fig. 3.8 for T_c and $U(0)$ in YBaCuO. A symmetrical pseudogap should also appear in the overdoping range, increasing with doping, as schematized by the T_f line in Fig. 4.13. The same model should apply for electron doped compounds, which indeed show experimentally a similar behaviour. To be more than a very qualitative description, such a picture should be elaborated from a better knowledge of the short range antiferromagnetism and its variation with doping and temperature. This would allow a more complete description of the wave functions and Fermi surface, especially in the underdoped range.

Finally, as stressed by the author, the superconducting gap has d wave symmetry in cuprates, as shown definitely by Tsuei and Kirtley in surface studies of triple points in polycristals as well as by Guy Deutscher in a number of beautiful tunnel and reflection experiments. These observations have been taken as proof for a direct electron–electron coupling of the electrons, as promoted e.g. by Pines. But if one takes into account the existing short range antiferromagnetism, the d gap can be explained by a usual phonon coupling with a supplementary (s or d) imaginary correction of the gap for overdoped samples, as indeed observed by Guy Deutscher.

In conclusion, this book progresses step by step in, what is in fact, a rather complex story. It introduces in a most simple way the concepts necessary for each chapter. It is clearly written and refers in many places to the personal experiences of the author. It is in a way aimed at the reflections of engineers about the present state and likely future of superconductivity. But it can benefit students who do not yet know enough about superconductivity, and can also be recommended to advanced researchers.

Jacques Friedel
Paris

Introduction

In the year 2011, the history of superconductivity will be one hundred years old. For a sub-field of physics, this is a very long period of time. One could have reasonably expected superconductivity to be by now a closed chapter of Physics. Yet, the discovery of the High T_c cuprates by Bednorz and Muller in 1986, showed that there remained more to be discovered and to be understood about superconductivity, than was expected after the consequences of the theory of Bardeen, Cooper and Schrieffer (BCS) had all been explored in depth.

It is the theme of these lecture notes that the cuprates belong to a class of superconductors which are, in many respects, fundamentally different from the conventional superconducting metals and alloys. The cuprates are not the only members of this class: granular superconductors, and at least some of the organic superconductors, discovered earlier, also belong to it, as well as other oxides.

Why do the cuprates have high critical temperatures? This is the question that a large number of researchers have been trying to answer for the last 19 years, so far with limited success only. They are close to an antiferromagnetic state, and this proximity is at the heart of some theories of high temperature superconductivity. But other oxides are not. The one property that all high temperature superconductors have in common, is that they are close to a metal-insulator transition. They are not metals, or alloys, in the ordinary sense: a very small change in composition is sufficient to transform them into insulators. Unfortunately, the enormously successful weak coupling BCS theory of superconductivity is really a theory that applies to metals only. It is not an accident that it did not predict high temperature superconductivity. It was even used by some to predict that high temperature superconductivity could not exist. In fact, it is probably a correct prediction that metals cannot be high temperature superconductors.

Our understanding of the new superconductors is still very imperfect, compared to that of metals and alloys. In these lecture notes, no attempt is made at reviewing the various mechanisms that have been proposed to explain High T_c. My purpose is more modest, and is limited at pointing out, on the basis of the experimental evidence, to the basic phenomenology of these materials. In many aspects of their behavior, they resemble more superfluid Helium than metal-superconductors. This suggests that electron pairs, bosons of zero total spin, may form above T_C — a feature that indeed cannot occur in metals. But there is still no indisputable proof that this is indeed the case.

Because of their high T_c, there has been a tremendous push towards the applications of the new superconductors, a somewhat hazardous enterprise in the absence of a good fundamental understanding of these fascinating materials. Yet, remarkable progress in the control of material quality, necessary for their application, has been achieved. It is also of great help in developing better fundamental experiments, which in turn have greatly improved our knowledge. There is now a substantial experimental basis for a condensation mechanism that is intermediate between BCS and Bose–Einstein. The consequences of this situation, particularly concerning the properties of the vortex state, have not yet been explored in depth. In this regard, we may look at granular superconductors as model systems, and draw on them to give us at least some intuition of the phenomenology of the cuprates.

A more fundamental question concerns the mechanism responsible for pair formation — quite a formidable task. Two schools of thought have been affronting each other. One of them holds that it is entirely due to electron-electron Coulomb interactions, known to be strong since the pristine cuprates are anti-ferromagnets, with the electron-phonon interaction playing no role at all in pair formation. The other holds that the electron-phonon interaction still plays an essential role in the cuprates. We shall try to summarize the main arguments of both schools. Here again, a comparison with granular superconductors proves instructive. Well before the discovery of the cuprates, it was shown that the critical temperature of granular Aluminum increases as Coulomb interactions are increased (here, by reducing the grain size which can by itself modify the electronic structure), up to the point where the metal-insulator transition is reached. This result flew in the face of the BCS–McMillan theory of superconductivity, in which a repulsive Coulomb interaction can only decrease the critical temperature. Evidently, this is a case where both the electron-phonon interaction, and electron-electron Coulomb interactions are at work. This unconventional situation is one of the reasons for reviewing in these notes some of the properties of granular superconductors.

But the main purpose of these notes is to outline what is really new about the cuprates from the standpoint of their phenomenology, and to provide a simple classification that can serve as a guide for practical applications.

Acknowledgments

Warm thanks are due to the group of current or former students at Tel Aviv University: Boaz Almog (whose photograph of a magnetically suspended superconducting wafer prepared by Mishael Azoulay is on the front cover), Roy Beck (whose help has been invaluable, particularly for preparing many of the figures and spending time in reading the manuscript), Yoram Dagan, Uri Dai, Gal Elhalal, Eli Farber, Nir Hass, Alexander Gerber, Meir Gershenson, Amir Kohen, Ralf Krupke, Guy Leibovitch, Zvi Ovadyahu, Yoash Shapira, Yoad Yagil — the contents of this book are in great part the result of common work and innumerable discussions with them. Without the expertise, teachings and friendly help of Enrique Grunbaum on electron microscopy none of the structural work on granular superconductors would have been done. The long standing collaboration on these materials with Peter Lindenfeld and the late W. McLean is also gratefully acknowledged. In more recent years, I have benefited from numerous discussions on High T_c with colleagues at the Heinrich

Hertz — Minerva Center on High T_c, and the Israel Science Foundation Center of Excellence on nanostructured materials, particularly Rudolf Huebener, Gad Koren, Oded Millo and Yosi Yeshurun.

The final two chapters reflect many exchanges with my colleagues in a number of review panels, notably that of the DOE program on superconductivity headed by James Daley, and at the workshops of the IEA Implementing Agreement on Superconductivity. I wish to thank them here for sharing with me some of their recent important results. Close cooperation with Yuli Milstein and with Alan Wolsky in the framework of this Agreement is kindly acknowledged.

It is clear that in between granular Aluminum and High T_c stands the momentous discovery of Georg Bednorz and Alex K. Mueller, with whom I have enjoyed 20 years of warm relations, and endless fruitful discussions in our quest for understanding these materials.

Finally, I wish to express my thanks to Jacques Friedel for accepting to write the foreword. I am very grateful to him and to Pierre Gilles de Gennes for their constant support, willingness to listen, comment, and in general, for giving me the benefit of their deep physical insight.

Contents

Chapter 1

Superfluidity

As their name indicates, superfluids flow without dissipation. This remarkable property, shared by systems as different as superconductors where the flowing particles are electrons, and superfluid Helium where there are atoms, is due to the macroscopic occupation of a quantum state. It is a manifestation of macroscopic quantum coherence, a notion fundamental to superfluidity.

1.1 The Landau critical velocity

Let us consider first the case of liquid Helium, and assume that its special properties below a critical temperature called the λ point are due to the fact that a macroscopic fraction of He atoms condense in a state of zero momentum, not subject to thermal agitation. Let us further assume that this macroscopic occupation is not immediately destroyed when we set the liquid in motion. All atoms initially at rest have now acquired a small velocity V. This is the velocity of the *condensate.* As we increase the velocity, we guess that at some point the condensate will start to be depleted by the creation of *excitations.* When will this happen?

An excitation is characterized by its momentum p, with respect to the frame of reference of the condensate, and by its energy $\epsilon(p)$. In the energetically most favorable case, the momentum of the excitation will be opposite to the velocity of the condensate, reducing the total kinetic energy. The change in energy of the system will then be, to first order in p,

$$\Delta E = \epsilon(p) - pV \tag{1.1}$$

This energy is positive as long as V is smaller than a critical velocity

$$V_c = \left[\frac{\epsilon(p)}{p}\right]_{\min} \tag{1.2}$$

This is the critical velocity introduced by Landau for the case of superfluid Helium. But it is, in fact, a very general expression, valid as well for superconductors.

In liquid Helium, the low lying excitations are phonons, or density waves, with an energy linear in p. The ratio $\left(\frac{\epsilon(p)}{p}\right)$ is then a constant. At higher momenta, there are other excitations, called *rotons*, giving a local minimum in $\epsilon(p)$, and therefore also a minimum in $\left(\frac{\epsilon(p)}{p}\right)$ (Fig. 1.1).

The dispersion curve $\epsilon(p)$ has been determined by neutron scattering experiments, so the measured value of V_c can be compared to that calculated from the Landau criterion. In fact, the experimental value is much smaller than the calculated one, presumably because of interactions between the superfluid and the walls of the channel in which it flows. However, a value very close to the theoretical one is found when ions are accelerated within the superfluid.

In superconductors, there are no low lying excitations near the Fermi level. The first excited states have an energy equal to the *energy gap* Δ (at least for the standard BCS superconductors) (Fig. 1.2). From Landau's expression, one then obtains for the critical velocity in superconductors:

$$V_c = \frac{\Delta}{p_F} \tag{1.3}$$

where p_F is the Fermi momentum. This is exactly the result given by the BCS theory. Note, however, that the existence of an energy gap is not a pre-requisite for the existence of superconductivity. All that is really needed is a macroscopic condensate. For instance, a finite critical velocity is still observed in *gapless superconductors,* obtained for instance by the introduction of magnetic impurities, or

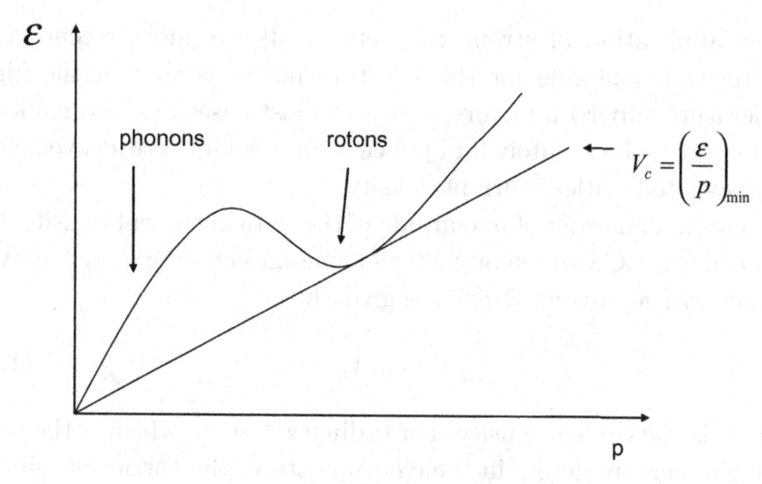

Figure 1.1: Excitation spectrum of superfluid Helium. Superfluidity does not require an energy gap, only a minimum value for (ϵ/p).

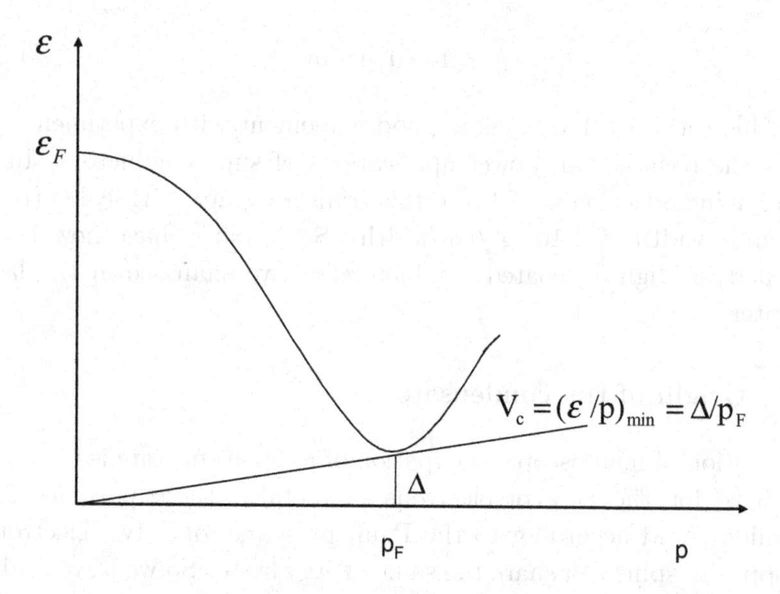

Figure 1.2: Electronic excitations in a superconductor. Landau's critical velocity is equal to the energy gap divided by the Fermi momentum. The excitation energy at zero momentum is the energy required to bring an electron from the bottom of the conduction band up to the Fermi level.

by the application of strong magnetic fields. A more recent, and very relevant example for these lecture notes, is that of the high temperature superconductors, which in most cases are also gapless, but do have — fortunately for applications — a large critical velocity, and associated critical current density.

To obtain an order of magnitude of the critical current density, let us consider a BCS superconductor having an energy gap of 1 meV.

The critical current density is given by:

$$j_c = neV_c \tag{1.4}$$

where n is the carrier density. For ordinary metals, which is the case of BCS superconductors in their normal state, the carrier density is of the order of $10^{22}/\text{cm}^3$. The Fermi wave vector is of the order of 1Å^{-1}, hence the Fermi momentum $hk_F \cong 1 \cdot 10^{-19}$erg·sec/cm. We obtain:

$$j_c \cong 1 \cdot 10^7 \text{A}/\text{cm}^2 \tag{1.5}$$

This theoretical value is in good agreement with experiment. It gives the basis for all power applications of superconductors. In a film having a thickness of 1 μm, this translates into a critical current per unit width of $1 \cdot 10^3$ A/cm-width. Such values have now been reached in High T_c coated conductors, as we shall see in the last chapter.

1.2 Origin of the condensate

The notion of macroscopic occupation of a quantum state is counter-intuitive for the case of electrons in metals: these particles are Fermions, and according to the Pauli principle, only two electrons of opposite spins can share the same energy state. So we leave aside, for the moment, superconducting metals, and consider the easier case of liquid Helium. Here, the particles are bosons and there is in principle no limit to the occupation number of a given state.

We assume, for simplicity, that there are no interactions between the Helium atoms. This idealized description is that of the *Bose Gas.*

We treat the Helium atoms as particle-waves having the de Broglie wave length:

$$\lambda = \frac{h}{p} \tag{1.6}$$

where p is determined by the thermal energy:

$$\frac{p^2}{2m} = k_B T \tag{1.7}$$

$$p = \sqrt{2mk_B T} \tag{1.8}$$

At low enough temperatures, the de Broglie wave length will eventually become larger than the inter-particle distance, hence a quantum description of the gas must be used. Treating the Helium atoms as free particle-waves in a box of size L, the density of states of this gas is given by:

$$N(E) = L^{-3}2^5\sqrt{2}\pi^4\left(\frac{m}{h^2}\right)E^{\frac{1}{2}} \tag{1.9}$$

and the occupation probability is:

$$f(E) = \frac{1}{\exp(\frac{E-\mu}{k_B T}) - 1} \tag{1.10}$$

where the value of the chemical potential μ is determined by the condition that the particle density n be kept constant:

$$n = \int_0^\infty N(E)f(E)dE \tag{1.11}$$

This equation sets the value of the chemical potential. Let us first notice that it must be negative, since otherwise, for small values of E, the occupation probability would be negative (Fig. 1.3). Since it cannot cross the zero value, there must be, below that temperature T_c, an increase in the low energy states occupation probability. The picture then is that below T_c, a macroscopic occupation of the state

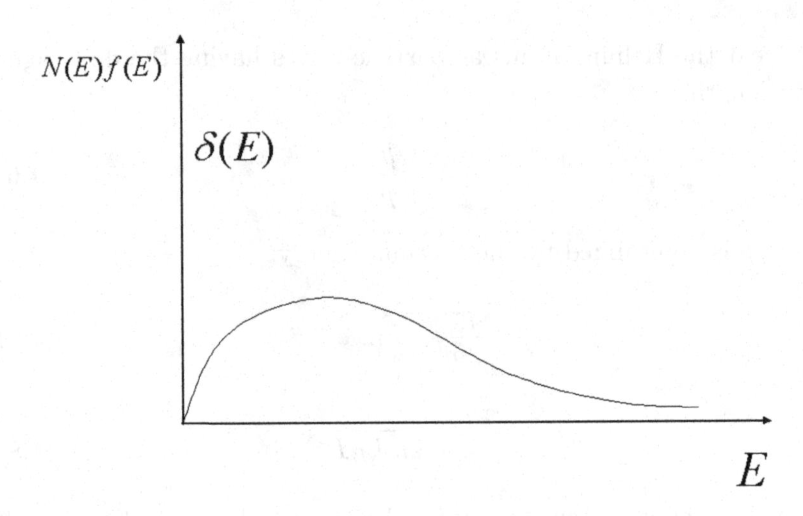

Figure 1.3: Below T_c, the number of particles in the zero energy state becomes macroscopic.

E=0 will start with all particles being eventually in that state at T=0.

We calculate the value of T_c as the temperature at which μ=0:

$$n = A\left(\frac{m}{h^2}\right)^{\frac{3}{2}} \int_0^\infty \frac{1}{\exp\left(\frac{E}{k_B T_c}\right) - 1} E^{\frac{1}{2}} dE \tag{1.12}$$

where A is a numerical constant. This gives:

$$T_c = 3.31\left(\frac{h^2}{m}\right) n^{\frac{2}{3}} \tag{1.13}$$

This expression has a simple physical interpretation: it is the temperature at which the de Broglie wave length of the particles becomes of the order of the inter-particle distance $a = n^{-\frac{1}{3}}$:

$$a \cong \frac{h}{\sqrt{2mk_B T_c}} \tag{1.14}$$

As soon as we cross T_c, the occupation of the zero energy state becomes macroscopic, and superfluidity appears. The superfluid fraction can move without dissipation with a finite velocity V, while the rest of the fluid (the "normal" fraction) stays at rest.

The remarkable fact is that the Bose–Einstein (BE) result for T_c has no adjustable parameter. If we put in the density of liquid Helium, we obtain a value of 3K, quite close to the experimental value of 2.14K. There is undoubtedly quite a bit of luck here, since the BE result neglects all interactions between the particles, an unreasonable assumption since the liquid is actually quite dense. Yet, the agreement is remarkable. One effect of the particle-particle interaction is that not all He atoms are condensed at T=0, but only approximately 10%. But this does not matter: what matters is that the condensed fraction is macroscopic.

1.3 Phase of the condensate

The quantum macroscopic condensate can be described by a wave function, having an amplitude and a phase:

$$\Psi = \mid \Psi \mid e^{i\varphi} \tag{1.15}$$

The amplitude of this wave function squared is the density of the condensed particles. In its ground state, i.e. when it is at rest, the phase is uniform and arbitrary. The importance of the phase appears in two kinds of situations: when we set the condensate in motion, and when we change the number of particles in the condensate.

When we set the condensate in motion, the current density of particles is proportional to the gradient of the phase:

$$\mathbf{j} = \frac{\hbar}{2mi}(\Psi^* \nabla \Psi - \Psi \nabla \Psi^*) \tag{1.16}$$

which we can interpret as the product of the density $\mid \Psi \mid^2$ by the velocity:

$$\mathbf{V} = \frac{\hbar \nabla \varphi}{m} \tag{1.17}$$

To Landau's critical velocity corresponds a critical phase gradient:

$$(\nabla \varphi)_c = \left(\frac{m}{\hbar}\right)\left(\frac{\epsilon(p)}{p}\right)_{\min} \tag{1.18}$$

Let us calculate the critical phase gradient in the case of a BCS superconductor having an energy gap Δ at $p = p_F$. We obtain:

$$(\nabla\varphi)_c = \frac{\Delta}{\hbar v_F} \tag{1.19}$$

where v_F is the Fermi velocity. The critical phase gradient is determined by a length scale:

$$(\nabla\varphi)_c = \pi\xi^{-1} \tag{1.20}$$

where:

$$\xi = \frac{\hbar v_F}{\pi\Delta} \tag{1.21}$$

This length scale is called the *coherence length*. For BCS superconductors, the critical current density is reached when the phase gradient is of the order of π divided by the coherence length. For a typical low temperature superconductor, having an energy gap of 1 meV and a Fermi velocity of $1 \cdot 10^8$cm/sec, it is of the order of 1000Å, several orders of magnitude larger than the interatomic distance.

The other case where the phase is important is that where we change the number of particles in the condensate. Suppose for instance that we introduce a weak link (a small orifice of atomic size) between two condensates, say two recipients filled with superfluid Helium. If there is no particle flow through this weak link, the phases of the two condensates must be identical. But suppose that we now set the superfluid in motion through this weak link. Then, a phase difference $\Delta\varphi$ will appear between the two condensates. By analogy with the case of a bulk current flow, we write:

$$\Delta\varphi = l\frac{m}{h}V \tag{1.22}$$

where l is the size of the orifice. In this experiment, the velocity V is equal to the flow of particles dN/dt, multiplied by the inter-particle distance, which we have assumed is of the same order as l. We finally find for the phase gradient:

$$\Delta\varphi = l^2\frac{m}{h}\frac{dN}{dt} \tag{1.23}$$

Since the phase difference cannot exceed π, without the system going back effectively to an equivalent state, this relation puts a limit to the flow of particles between the two condensates that can take place *without dissipation.* It is one aspect of the Josephson effect. Of course, this *gedanken* experiment is highly impractical: we assumed that the orifice was of atomic size, otherwise the opening would be a regular channel and dissipation-less flow would just be identical to bulk superfluidity. Not only is such a small orifice difficult to make in a controlled way, but the outcome of the experiment becomes in this limit very sensitive to the boundary condition for the wave function at the walls, on the atomic scale. We shall have the occasion to return later to this boundary condition problem.

Josephson effects are in fact much better studied in superconductors. In that case, the weak link can be a well controlled tunnel junction, or a constriction of size ξ, which as we have shown is much larger than the interatomic distance. What is important to note at this stage, is that dissipation-free currents can flow through weak links between superfluid condensates.

1.4 Two-Fermion superfluids

Now let us go back to the case of electrons in solids. As already noted, electrons will not condense as Helium atoms do, because they are Fermions. But let us imagine that there exists a strong attractive potential between electrons, so that electrons can form pairs. This might happen for instance if there is a strong electron-phonon interaction, in which case electrons can dig their own potential well in the lattice to form polarons, and under certain circumstances, bipolarons. We assume that this potential is strong enough to produce pair formation at high temperature, so that at low temperatures the pairs are very stable and can be considered as bosons. In such a case, Bose–Einstein condensation of these pairs will occur as we have calculated: to obtain the transition temperature, we only need to replace the Helium atom mass by twice the electron mass:

$$T_{co} = 3.31 \frac{h^2}{2m_e^*} \left(\frac{n}{2}\right)^{\frac{2}{3}} \tag{1.24}$$

where n is the electron density, and m_e^* is the effective mass of the electrons.

Because the electron mass is very much smaller than that of He atoms, this expression predicts for our hypothetical two-Fermion superfluid a very high condensation temperature.

How realistic is such a picture? Several stringent conditions must apply for such a pair formation, and pair condensation, to occur:

(i) the binding energy ϵ_0 of the pair must be of the order of, or larger than the typical energy that free (not bound) electrons would have, namely the Fermi energy:

$$\frac{\epsilon_0}{2} > E_F \tag{1.25}$$

(ii) the formed pairs must be mobile in order for them to condense.

The first condition certainly cannot be met in ordinary metals, where the Fermi energy is large (on the order of several eV), and screening is strong, preventing the formation of strong attractive potentials between electrons. On the other hand, if screening is weak and interactions are strong, we may be dealing with an insulator rather than with a metal, and pairs — if they are indeed formed — will tend to be localized, and the second condition will not be met.

Our conclusion is that, if at all, these two conditions can only be met in a material that is very close to a metal-insulator transition. Then, the free carrier density is small, the Fermi energy is reduced compared to that of an ordinary metal, interactions are stronger, but electrons are not yet strongly localized. Keeping these caveat in mind, we see that the value of the BE T_c is not quite as high as we might naively have expected from the ratio of the masses: instead of a typical metallic density of the order of $1{\cdot}10^{22}/\text{cm}^3$, we must rather use a value typical of a degenerate semiconductor, say $1{\cdot}10^{21}/\text{cm}^3$; and the effective mass of the electron may be substantially larger than that of the free particle, because we are in a regime of strong interactions. Taking both effects into account, the BE T_c would only be one to two orders of magnitude larger than that of liquid He, instead of four. But this is still a very high condensation temperature, in the 100K range.

Superconductivity is indeed observed in a number of systems that are close to the metal-insulator transition: granular superconductors have often critical temperatures higher than that of the parent bulk superconductor; organic superconductors and last but not least, the high-temperature superconductors cuprates. These are natural candidates as Bose–Einstein superconductors.

Assuming that such pair formation and condensation are possible, the normal state is characterized by the amplitude of the potential, V_m and by its range a. For pairs to form at all, we must have:

$$V_m > \frac{\hbar^2}{2ma^2} = V_a \tag{1.26}$$

We wish to calculate the pair binding energy ϵ_0 and its extension in the potential well, a_0. In the strong coupling limit $V_m \gg V_a$, the pair is strongly bound at the bottom of the potential well, and its binding energy is close to the depth of the well

$$\epsilon_0 \approx V_m \tag{1.27}$$

For a deep potential well we can use the harmonic approximation. The pair extension is then given by the harmonic oscillator solution, it varies as the inverse fourth power of the potential amplitude, i.e. of the pair binding energy

$$a_0 \propto \epsilon_0^{-1/4} \tag{1.28}$$

In the normal state, there is a gap:

$$\Delta_p = \frac{\epsilon_0}{2} \tag{1.29}$$

Δ_p is the *single particle excitation energy* of the system. If, for instance, we try to inject a single electron in our material from a normal metal through a tunnel junction, we shall have to apply a potential difference across the junction of at least Δ_p/e.

Electrons are paired over the length a_0. This is in a sense the equivalent of the *coherence length* that we introduced earlier for the case of a BCS superconductor. But there is of course a major difference: in the present case, the gap and the pairing length are properties of the *normal state*. When the Bose–Einstein condensation of the pairs occurs at some lower temperature, there will be no change in this gap. In the condensed state, a new length scale appears, which characterizes the scale over which the phase of the condensate (and not its amplitude) can vary. This is really the coherence length of the Bose–Einstein condensate. Neither the pairing length, nor the coherence length are anymore related to the single particle energy gap in the way they are in a BCS superconductor.

We conclude that, if Bose–Einstein superconductivity does exist, it will be characterized by a large tunneling gap, only weakly temperature dependent on the scale of the condensation temperature, by a small pairing length, of the order of interatomic distances, also temperature independent and distinct from the length scale that characterizes changes in the phase of the condensate. In principle, these properties are easily distinguishable from those of BCS superconductors in which the gap is a thermodynamical gap, proportional to T_c at low temperatures, and going to zero at T_c, and the pairing (in that case, also coherence) length is large compared to interatomic distances, temperature dependent (it diverges at T_c) and given by the Fermi velocity divided by the gap at T=0.

1.4.1 The Meissner effect

In a two-Fermion superfluid, the expression for the current that we have used for non-charged superfluids, Eq. (1.16), must be modified to take into account the charge 2e of our new "two-Fermion bosons" . By analogy with the expression for the current for particles of charge 2e and wave function $\Psi(r)$, we write:

$$\mathbf{j} = \frac{e\hbar}{im^*}(\Psi^* \nabla \Psi - \Psi \nabla \Psi^*) - \frac{4e^2}{m^* c} \mid \Psi \mid^2 \mathbf{A} \qquad (1.30)$$

where we have taken the particles to be of mass m^*, and $\mathbf{A}$ is the vector potential.

The vector potential will be non-zero in the presence of applied magnetic fields or currents, or both. This covers a wide variety of situations. In general, Ψ and $\mathbf{A}$ will be position dependent, and both dependencies must be known in order to calculate the current $\mathbf{j}(\mathbf{r})$.

The situation is much simpler when the applied fields and currents are small: we can then assume that $\mid \Psi \mid$ retains its zero-field value, $\mid \Psi_0 \mid$. Taking the **curl** of both sides, and making use of:

$$\mathrm{curl}\,\mathbf{h} = \frac{4\pi\mathbf{j}}{c} \tag{1.31}$$

we obtain:

$$\mathrm{curlcurl}\,\mathbf{h} = \frac{16\pi e^2}{m^* c^2} \mid \Psi_0 \mid^2 \mathbf{h} \tag{1.32}$$

This relation defines a length scale λ such that:

$$\lambda^{-2} = \frac{16\pi e^2}{m^* c^2} \mid \Psi_0 \mid^2 \tag{1.33}$$

To take a specific example, assume that we have applied an external magnetic field H parallel to the surface of a very thick sample. The equation for the current then tells us that it will decay exponentially inside the sample, over the length scale λ. So will the internal magnetic field h. If we orient the x axis normal to the surface of the sample, this internal field will obey the relation:

$$h(x) = H \exp -\left(\frac{x}{\lambda}\right) \tag{1.34}$$

since the parallel component of the field must be continuous across the surface of the sample.

The assumption of a rigid, macroscopic wave function to describe the condensate was originally proposed by London. λ is called the London penetration depth. The exact nature of the condensate — whether it is a Bose–Einstein condensate as we discuss here, or a more classical BCS condensate of which we recall the main properties in a later chapter — does not affect the validity of the London equation.

1.4.2 Flux quantization

Let us consider a hollow cylinder whose wall is much thicker than the penetration depth λ. Inside the wall we can find a path around the cylinder along which the local magnetic field and the current is zero, according to Eq. (1.34). Integrating Eq. (1.30) along such a path, and using Eq. (1.15), we obtain:

$$\oint \nabla\varphi \cdot d\mathbf{l} = \frac{2e}{\hbar c}\oint \mathbf{A}.d\mathbf{l} \tag{1.35}$$

Since the l.h.s must be a multiple of 2π, the flux enclosed within the selected path is given by:

$$\Phi = n\Phi_0 \tag{1.36}$$

where n is an integer and:

$$\Phi_0 = \frac{hc}{2e} \tag{1.37}$$

is the flux quantum.

To go back to our Bose–Einstein notation, we replace $\mid \Psi_0 \mid^2$ by the superfluid density n_s=$n/2$, and m^* by $2m$. Assuming that all particles have condensed at T=0, we have:

$$\lambda(0)^{-2} = \frac{4\pi e^2}{c^2}\frac{n}{m} \tag{1.38}$$

Now, we remember that the Bose–Einstein condensation temperature is a power law of the particle density (Eq. 1.13), here $(\frac{n}{2})$. Therefore, in our two-Fermion superfluid, the condensation temperature is a power law of the zero-temperature penetration depth:

$$k_B T_c \propto 3.31\left(\frac{c^2}{4\pi e^2}\right)^{\frac{2}{3}} \frac{h^2}{2m^{\frac{1}{3}}}\lambda(0)^{-\frac{4}{3}} \tag{1.39}$$

This relation can in principle be checked quantitatively. Note however that the effective mass of the individual electrons must be determined independently. A power law relation between T_c and $\lambda(0)$ ($T_c \propto \lambda(0)^{-2}$) has indeed been discovered by Uemura, and interpreted as an indication that the superfluid condensation in these materials is of the Bose–Einstein type. This will be discussed in more detail in the next chapter.

1.5 BCS superconducting metals

The case of the two-Fermion Bose–Einstein superfluid is a very special one. As we have seen, it can only occur in a material that is close to a metal-insulator transition. In fact, it is a well established fact that most superconductors are well behaved metals, with strong electronic screening and relatively weak interactions, so that this description cannot be applied to them.

Yet, the principle that the condensed state, characterized by a macroscopic occupation of a quantum state, is composed of pairs of electrons of opposite spins must apply.

1.5.1 Condensation energy

The microscopic description of metal-superconductors starts from the existence of a large Fermi surface, and correspondingly large Fermi energy, namely much larger than the energy scale — the energy gap Δ — that characterizes the superconducting state. The difference between the electronic structure of the normal and superconducting state is then seen only close to the Fermi level. Only electrons within an energy Δ from the Fermi level, have their energy substantially lowered in the condensed state; hence, the condensation energy per unit volume is of the order of $N(0)\Delta^2$ (the exact expression is $\frac{1}{2}N(0)\Delta^2$) where $N(0)$ is the normal state density of states, and the condensation energy per particle is of the order of $(\frac{\Delta^2}{E_F})$. Because this energy is so small, we might think that superconductivity in metals is going to be very fragile. In fact, this is not the case: perturbations of the superconducting state (due for instance to thermodynamical fluctuations) will take place over the scale of the coherence length, because this is the length scale over which the Ψ function can vary (as we have seen in the case of its gradient, Eq. (1.19)). Because the coherence length in a BCS superconductor is very large as we have seen, not just one pair will be affected, but many. The relevant energy scale for these perturbations is not the condensation energy per particle, *but the condensation energy per coherence volume:*

$$U \cong N(0)\Delta^2\xi^3 \tag{1.40}$$

Using Eq. (1.21) for ξ and the free electron expression for the normal state density of states, we get:

$$U = \frac{2}{\pi^5}\frac{E_F^2}{\Delta} \tag{1.41}$$

This is a very large energy — much larger even than the Fermi energy in the considered situation $E_F \gg \Delta$. The condensate is in fact very robust. We say that coherence in metal-superconductors is very strong, a useful property that has important practical consequences for their applications.

In a BCS superconductor, it is the destruction of pairs by the thermal energy that determines the critical temperature. Vice versa, in metal-superconductors, pairs form and condense at the same temperature T_c. In the weak coupling limit, $\Delta \ll E_F$, the critical temperature is proportional to the gap, $2\Delta = 3.5 k_B T_c$.

1.5.2 The BCS wave function

BCS introduced the following wave function to describe the condensed state:

$$\widetilde{\Phi} = \prod_k (u_k + v_k a_{k\uparrow}^+ a_{-k\downarrow}^+)\Phi_0 \tag{1.42}$$

where Φ_0 describes the vacuum, the operator $a_{k\uparrow}^+$ creates an electron of spin up and wave vector k and the operator $a_{-k\downarrow}^+$ an electron of spin down and wave vector $-k$. The normalization of $\widetilde{\Phi}$ is ensured by the condition $u_k^2 + v_k^2 = 1$. The probability to find a condensed electron at wave vector k and spin α is given by:

$$v_k^2 = \langle \widetilde{\Phi} \mid a_{k\alpha}^+ a_{k\alpha} \mid \widetilde{\Phi} \rangle \tag{1.43}$$

where the operator $a_{k\alpha}$ destroys a condensed electron in the state $k\alpha$.

For an isotropic gap Δ the energy of an electron of wave vector k in the condensed state is given by:

$$\epsilon_k = \sqrt{\xi_k^2 + \Delta^2} \tag{1.44}$$

where ξ_k is the energy in the normal state of an electron of wave vector k, measured from the Fermi level, and:

$$v_k^2 = \frac{1}{2}\left(1 - \frac{\xi_k}{\sqrt{\xi_k^2 + \Delta^2}}\right) \tag{1.45}$$

Electrons in the condensed state are spread over an energy of the order of Δ around the Fermi surface. They lie both above and below the Fermi energy, while in the normal state at zero temperature there are no electrons above the Fermi wave vector. Hence the average kinetic energy of the electrons in the condensed state is *larger* than that in the normal state. The energy gain in the condensed state comes about because the lowering of the potential energy is larger than the increase of the kinetic energy. This is a fundamental difference with Bose–Einstein condensation, where in the absence of interaction there is no change in the potential energy, and the gain in energy in the condensed state comes about entirely from a diminution of the kinetic energy. Since the kinetic energy of a boson is of the order of k_BT_c in the normal state at the transition, and zero in the condensed state, this energy gain is of the order of k_BT_c per boson, or nk_BT_c per unit volume, where n is the boson density.

There are thus two important differences between the two modes of condensation:

(i) In the BCS condensation mode, the kinetic energy increases, while it decreases in the BE condensation. A measurement of the *sign* of the change in the kinetic energy upon condensation is of particular interest when there are indications that pairing may occur at higher temperatures. We come back later to this point in Chapter 7 when we discuss experimental ways of determining the actual mode of condensation in the cuprates.
(ii) The condensation energy per boson is much larger in BE condensation than in BCS condensation. The former is of the order of k_BT_c, and the later of the order of $\frac{\Delta^2}{E_F}$ or within numerical factors $k_BT_c\left(\frac{T_c}{T_F}\right)$, where T_F is the Fermi temperature, $T_F \gg T_c$. One may be tempted to conclude that the BE condensed state is "stronger"

(for a given transition temperature) than the BCS state. This is in fact wrong: the BCS state is more rigid against fluctuations than the BE state. The reason is that the later has a much shorter coherence length (of the order of the inter-boson distance) than the former, and what actually determines fluctuation amplitudes is the *condensation energy per coherence volume,* rather than the condensation energy per boson. For the BE condensate, the condensation energy per coherence volume is then of the order of k_BT_c, while it is as we have seen of the order of $\left(E_F^2/k_BT_c\right)$ in a BCS superconductor.

These considerations are not purely academic. Fluctuations can have a strong impact on some important practical properties, such as the ability to retain a zero resistance state in the presence of strong magnetic fields and strong currents. There is a close connection between the mode of condensation and applications. It will be discussed in some details when we review properties of the vortex state in the last two chapters.

While the two kinds of condensation that we have described are physically very different, Eagles, Leggett and Nozieres and Schmitt-Rink (see further reading for references) have shown that as the interaction strength is increased, the transition between the two modes is in fact continuous as a function of the ratio of the pair breaking energy to the chemical potential.

1.6 Summary

Superfluids can be divided broadly into two categories:

Short coherence length superfluids

They include superfluid Helium, and possibly electronic systems near a metal-insulator transition called two-fermion superconductors, in which electron pairs would form above the condensation temperature. Their main properties are:

- a coherence length of the order of some microscopic scale, independent of the condensation temperature T_c.

- a condensation energy per particle, hence also per coherence volume, of the order of k_BT_c.

- a temperature independent gap for the creation of single particle excitations, but no thermodynamical gap.
- a penetration depth related to the condensation temperature by a power law.

Long coherence length superconductors

They include superconducting metals and alloys. Their main properties are:

- a coherence length much larger than interatomic distances, inversely proportional to T_c.
- a condensation energy per particle much smaller than $k_B T_c$, but a condensation energy per coherence volume much larger than $k_B T_c$.
- a thermodynamical gap in the excitation spectrum.
- a penetration depth unrelated to T_c.

1.7 Further reading

For a description of the BCS condensation, see P.G. de Gennes, "Superconductivity of Metals and Alloys", Benjamin Inc., New York 1966.

For one of the original articles on the cross over between the Bose–Einstein and the BCS condensation modes, see P. Nozieres and S. Schmitt-Rink, J. Low Temp. Phys. **59**, 195 (1985). For a recent review on this topic, see for instance Q. Chen *et al.*, Phys. Reports **412**, 1 (2005) and references therein.

Chapter 2

Coherence length, penetration depth and critical temperature

We have introduced in the preceding chapter the notion of a coherence length, ξ, characteristic of the condensate, as the minimum length scale over which the phase of the order parameter can be varied by an angle of the order of π, without breaking up the condensate. Assuming that there is an energy gap Δ at $p = p_F$ we found that:

$$\xi = \frac{\hbar v_F}{\pi \Delta} \tag{2.1}$$

This expression is in fact only valid when the normal state mean free path is longer than ξ (clean limit).

On the other hand, based on the analogy with a Bose–Einstein condensate, we introduced the notion that in a superconducting metal the condensate is composed of pairs of electrons. These electrons have opposite spins and opposite momenta, so that their center of mass is at rest in the ground state. We now need to clarify how this notion of paired electrons is related to Eq. (2.1). This is the object of the first section of this chapter.

We then turn to the experimental verification of this relation. If condensation is of the BCS type, where pairs form and condense simultaneously at the critical temperature T_c, there is only one single energy scale in the superconducting state. The gap and T_c are then proportional to each other. Because the Fermi velocity does not vary much amongst metals, Eq. (1) then implies that *the product of the critical temperature by the coherence length must be approximately the same for all superconductors in which condensation is of the BCS type.*

To check whether this coherence law applies, we shall need to discuss experimental methods that are used for the determination of the coherence length. These methods are based on the Ginzburg–Landau (GL) theory.

We shall find out that the BCS coherence law does indeed apply to superconducting metals, but does not to a number of other superconductors, such as high T_c cuprates. One may then question whether condensation in these materials is of the BCS type. At the end of this chapter, we shall therefore return to the case where they are pre-formed pairs. As we have shown in the previous chapter, the critical temperature is then related to the penetration depth rather than to the coherence length. Indeed, this is what happens in the cuprates. We emphasize that the question of the nature of the condensation — BCS or non-BCS — is distinct from that of the nature of the interactions that leads to the formation of pairs. This is a separate issue, that will be dealt with in Chapter 8 where we examine different kinds of interactions that can lead to this formation.

2.1 Origin of the coherence length in superconducting metals

One way to probe the condensate is to produce single particle excitations above the ground state. For instance, one can irradiate the superconductor with photons, and determine the frequency above which excited states are produced. Since we have assumed that the condensate is composed of pairs, these excited states must be the result of pair breaking, which occurs when:

$$h\nu > E \tag{2.2}$$

where E is the binding energy of the pair. The pair breaking energy per excitation is then:

$$\Delta = \frac{E}{2} \tag{2.3}$$

The macroscopic wave function that describes the condensate is the product of the wave functions of all the pairs. As first proposed by Cooper, we isolate somewhat arbitrarily one of them, composed of electrons located at points $\mathbf{r}_1$and $\mathbf{r}_2$, and write its wave function:

$$\psi(\mathbf{r}_1 - \mathbf{r}_2) = \sum_k g(k) e^{i\mathbf{k}\cdot(\mathbf{r}_1 - \mathbf{r}_2)} \tag{2.4}$$

Assuming that $\Delta \ll E_F$, the sum extends over wave vectors near the Fermi value k_F, in a range:

$$\delta k \approx \left(\frac{d\xi}{dk}\right)^{-1}_{\xi=0} .\Delta \tag{2.5}$$

where $\xi(k)$ is the single particle energy, counted from the Fermi level, in the normal state. Since:

$$v_F = \frac{1}{\hbar}\left(\frac{d\xi}{dk}\right)_{\xi=0} \tag{2.6}$$

we have:

$$\delta k \approx \frac{\Delta}{\hbar v_F} \tag{2.7}$$

From the Schrodinger equation for the pair wave function:

$$-\frac{\hbar^2}{2m}(\nabla_1^2 + \nabla_2^2)\psi + V(\mathbf{r}_1, \mathbf{r}_2)\psi = (E + 2E_F)\psi \tag{2.8}$$

we obtain an equation for the $g(k)'s$:

$$\frac{\hbar^2}{m} k^2 g(k) + \sum_{k'} g(k') V_{k,k'} = (E + 2E_F) g(k) \tag{2.9}$$

where:

$$V_{k,k'} = \frac{1}{L^3}\int V(\mathbf{r}) e^{i(\mathbf{k}-\mathbf{k}').\mathbf{r}} dr \tag{2.10}$$

Making the assumption that $V_{k,k'}$ is a constant for $| \mathbf{k} - \mathbf{k}' | < \delta k$, and zero elsewhere:

$$\sum_{k'} g(k')V_{k,k'} = C \tag{2.11}$$

and we obtain:

$$g(k) = \frac{C}{E - 2\xi(k)} \tag{2.12}$$

Here we do not attempt to calculate E, but assume that it has been determined experimentally, for instance by the irradiation experiment mentioned above.

We can now calculate the mean square distance between the two members of the pair, ρ:

$$\rho^2 = \frac{\int |\psi(\mathbf{r}_1 - \mathbf{r}_2)|^2 R^2 d\mathbf{R}}{\int |\psi(\mathbf{r}_1 - \mathbf{r}_2)|^2 d\mathbf{R}} \tag{2.13}$$

$$\rho^2 = \frac{\sum |\nabla_k g(k)|^2}{k \sum |g(k)|^2} \tag{2.14}$$

$$\rho^2 = \frac{(\hbar v_F)^2 \int d\xi_k \left(\frac{\partial g}{\partial \xi}\right)^2}{\int g^2 d\xi_k} \tag{2.15}$$

where we have taken into account that the major contribution to the integral comes from the vicinity of the Fermi energy, where the density of states is constant and we can replace $d\xi/dk$ by $\hbar v_F$. The exact result depends somewhat on the limits of integration. In the original Cooper problem where two electrons are added to the Fermi sea, $g(k) \equiv 0$ for $k < k_F$ and the integral runs from 0 to ∞. Then:

$$\rho = \frac{2\hbar v_F}{\sqrt{3}|E|} \tag{2.16}$$

The main point is that we obtain a length scale which is similar to that given by Eq. (2.1): the physical meaning of ξ is that it is the average size of a pair having the binding energy E. It is therefore natural that one cannot modify significantly the wave function over a length scale shorter than that size, without breaking the pairs that constitute the condensate. Note that the above calculation was made under the assumption that there is a large Fermi sea, $E \ll E_F$. Under this condition, it is easily checked from Eq. (2.16) that $\rho \gg k_F^{-1}$, or $\rho \gg a$ where a is the interatomic spacing. This calculation is only valid for long coherence length superconductors. The large Fermi sea plays a crucial role in the condensation when the electron-electron interaction is weak, as is the case in metals, in contrast with the case where the interaction is strong and it is pre-formed pairs that condense (BE condensation).

2.2 Experimental methods for the determination of the coherence length

If the superfluid density is small, it is possible to write the free energy of the condensate as a series development of that density. In their theory of superconductivity, Ginzburg and Landau (GL) introduced in that development an additional term that takes into account possible variations in space of the condensate wave function — both its amplitude and its phase:

$$F_s(\psi) = F_n + a|\psi|^2 + \frac{b}{2}|\psi|^4 + \frac{1}{m^*}\left|\left(- i\hbar\mathbf{\nabla} - \frac{2e\mathbf{A}}{c}\right)\psi\right|^2 \tag{2.17}$$

This development is based on the experimental evidence that the transition at T_c is a second order one. In their original paper GL used a charge e^* thus leaving open the possibility that it could be different from the charge of one electron. They had however no idea that the charge should be that of electron pairs, as seems natural to us today when we think of the condensation as that of bosons, see the section on "Two-Fermion superfluids" in Chapter 1. As for the mass used here, it has no special meaning (since the relation of ψ with the electrons density is not specified) except that as shown by GL Eq. (1.30) can be obtained by minimizing Eq. (2.7) with respect to small variations of **A**, and in order to be consistent with the expression of the London penetration depth the mass should be equal to twice the electron mass.

In order to ensure that the equilibrium value of ψ will be zero above T_c, GL takes:

$$\begin{aligned} a &= \overline{a}(T - T_c) \\ b &= \text{constant} \end{aligned} \tag{2.18}$$

so that in the absence of an applied vector potential, the equilibrium value of $|\psi|^2$,

$$|\psi_0|^2 = -\frac{a}{b} \tag{2.19}$$

goes linearly to zero at T_c. In the GL theory, the amplitude of the pair wave function is zero above T_c. In the absence of an applied vector potential **A**, the phase is uniform: phase coherence and pair amplitude appear together at T_c. In other words, the formation of pairs and the macroscopic condensate are assumed to be simultaneous. This is indeed what happens in the BCS condensation, as we have seen. As said in Chapter 1, other possibilities exist. At the end of this chapter, we shall specifically consider a case — that of granular superconductors — for which a modified version of the GL free energy allows to treat a case where a finite pairing amplitude exists before coherence is achieved.

For the case of a BCS condensation, we shall now consider two situations where the value of the coherence length determines the (depressed) temperature at which nucleation of the superconducting state takes place. A measurement of that temperature allows an experimental determination of the coherence length.

2.2.1 Boundary effect

We consider a slab of a superconducting metal of thickness d_s sandwiched between a dielectric (such as vacuum) on one side and a strong pair breaking material, such as a ferromagnetic metal on the

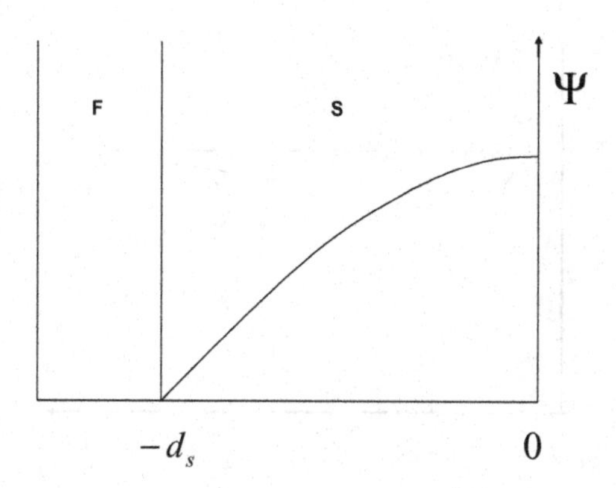

Figure 2.1: A superconducting slab is bounded on one side by a strong pair breaking layer, and on the other side by dielectric. The imposed spatial variation of the order parameter raises the free energy of the superconducting state and lowers the critical temperature.

other side (for instance, a strong exchange field acting on the opposite spins of the pairs will break them up). On the dielectric bound surface, we take $\nabla_{\perp}\psi$ =0, and on the magnetic side, ψ =0 (Fig. 2.1).

By forcing the ψ function to bend over the thickness d_s, we add a $|\nabla\psi|^2$ term to the free energy: at $T = T_c$, $F_s(\psi)$ will still be positive, the condensate will not appear. To render the slab superconducting, we must lower the temperature further ($|a|$ becomes larger). To calculate the new critical temperature, we first write that in any case, $F_s(\psi)$ as given by Eq. (2.14) with **A**=0 must be at a minimum with respect to small variations of ψ:

$$a\psi + b|\psi|^2\psi - \frac{\hbar^2}{2m}\frac{d^2\psi}{dx^2} = 0 \tag{2.20}$$

When nucleation occurs, ψ is infinitesimally small, we can neglect the cubic term and obtain the solution:

$$\psi = \psi \cos\left(\frac{x}{\xi(T)}\right) \tag{2.21}$$

where we have introduced the temperature dependent coherence length:

$$\xi^2(T) = \frac{\hbar^2}{2m|a|} \tag{2.22}$$

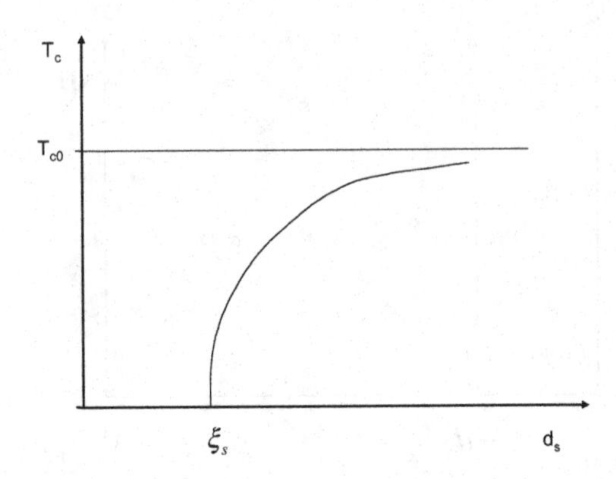

Figure 2.2: Schematic variation of the critical temperature of the F/S bilayer represented in Fig. 2.1. $\xi_S = \frac{\pi}{2}\xi(T=0)$

The solution for ψ must satisfy the condition ψ =0 at $x = d_s$. This gives us an implicit equation for the temperature of nucleation:

$$d_s = \frac{\pi}{2}\xi(T) \tag{2.23}$$

By measuring $T_c(d_s)$ we can directly determine experimentally $\xi(T)$. The coherence length diverges at the critical temperature of the bulk material T_{c0}, reached asymptotically as $d_s \to \infty$. The value of d_s for which T_c reaches zero is a measure of the zero temperature coherence length (Fig. 2.2).

2.2.2 Nucleation field

We now consider a situation where superconductivity has been quenched by the application of a strong field, which is then lowered progressively. We wish to calculate the field value at which superconductivity will nucleate. In the presence of a vector potential, minimization of the free energy with respect to small variations of ψ leads to:

$$a|\psi| + b|\psi|^2\psi + \frac{1}{2m}\left(- i\hbar\mathbf{\nabla} - \frac{2e\mathbf{A}}{c}\right)^2 \psi = 0 \tag{2.24}$$

As in the preceding case, ψ will be infinitesimally small when nucleation will occur. One can neglect the cubic term. In addition, according to Eq. (1.28), superfluid currents will also be infinitesimally small, therefore the field inside the superconductor will be equal to the applied field. The equation to be solved:

$$\frac{1}{2m}\left(-i\hbar\boldsymbol{\nabla}-\frac{2e\mathbf{A}}{c}\right)^2\psi=-a\psi \tag{2.25}$$

is then the same as that of a particle of charge $2e$ and mass m placed in a uniform magnetic field. The solution of lowest energy (equal to $-a$) corresponds to the circular motion of that particle, in a plane perpendicular to the applied field, with the cyclotron frequency $\omega_c=\frac{2eH}{mc}$:

$$\begin{aligned}-a&=\frac{1}{2}\hbar\omega_c\\ -a&=\frac{\hbar eH}{mc}\end{aligned} \tag{2.26}$$

It is again convenient to write this result in terms of the coherence length:

$$H_n(T)=\frac{\Phi_0}{2\pi\xi^2(T)} \tag{2.27}$$

where Φ_0 is the flux quantum (Eq. 1.37). By measuring the field at which superconductivity nucleates at temperature T, we can determine $\xi(T)$. Equation (2.27) together with Eq. (2.22) predict that the nucleation field is proportional to $|a|$. Its linear variation with temperature constitutes a verification of the GL choice for that coefficient. This linear variation is well verified for superconducting low T_c metals.

2.2.3 Nucleation field and thermodynamical critical field

The nucleation field is in general distinct from the field at which the free energy of the superconducting state is equal to that of the normal state, assuming that in the superconducting state the field does not penetrate further than the London penetration depth (Meissner state). This field is called the thermodynamical critical field, H_c. At

H_c, the magnetic energy $\frac{H_c^2}{8\pi}$ is equal to the condensation energy, ΔF. Assuming that the amplitude of ψ is not modified by the applied field (the rigid wave function hypothesis of London):

$$\Delta F = \frac{a^2}{2b} \tag{2.28}$$

Putting together Eqs. (1.33), (2.16) and (2.19), we obtain:

$$H_c = \frac{\Phi_0}{2\pi\sqrt{2}\lambda\xi} \tag{2.29}$$

Comparing the expressions for H_c and for H_n we obtain:

$$H_n = H_c\kappa\sqrt{2} \tag{2.30}$$

where:

$$\kappa = \frac{\lambda}{\xi} \tag{2.31}$$

is the GL parameter that characterizes the superconductor.

There are two distinct kinds of nucleation:

(i) if $\kappa > 1/\sqrt{2}$, $H_n > H_c$. *Type II superconductors*

Nucleation occurs at a field where the Meissner phase is not stable. The phase that nucleates is called the mixed phase or vortex phase, it contains both superconducting regions from which the magnetic field is excluded, and regions where it penetrates. As we have remarked in Chapter 1, the flux in these regions, called vortices, must be quantized in units of Φ_0. On opposite sides of the center of these vortices, the phase of ψ differs by π. Because the phase gradient must be finite — it cannot exceed $\frac{\pi}{\xi}$ — the amplitude of ψ must be zero at the center of the vortex. It will recover over the length ξ, as in the boundary problem that we have considered above (Fig. 2.3). The region of size ξ where the amplitude of ψ is depressed is called the core of the vortex, while the currents around the core circulate within a radius λ.

H_n is the field up to which the vortex phase is stable. From Eq. (2.27), we see that at that field the vortex cores basically overlap,

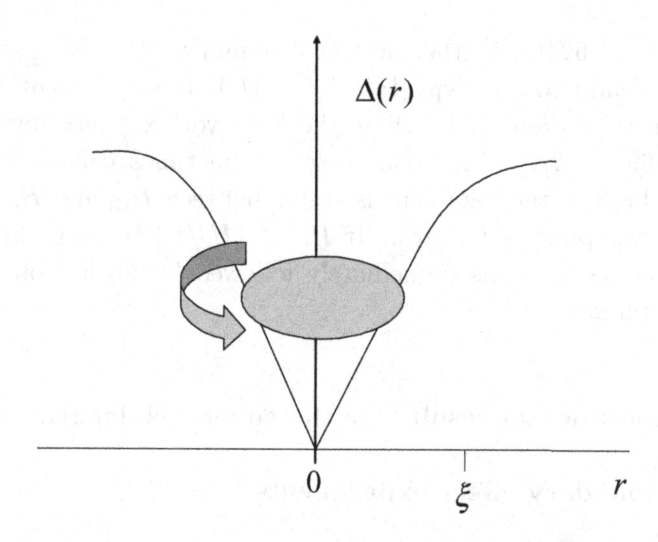

Figure 2.3: Circulating currents around a vortex impose a phase difference of π between opposite points on a circle surrounding the center. At distances of the order of ξ or less, the strong phase gradient imposes a reduction of the order parameter, down to zero at the center.

each carrying a flux quantum. The nucleation field is also the upper critical field H_{c2}.

(ii) $\kappa < 1/\sqrt{2}$, $H_n < H_c$. *Type I superconductors*

At H_n, the Meissner phase is the stable one. H_n is now a supercooling field, the lowest field down to which the normal phase remains metastable: down to H_n, there is still a local minimum of the free energy at ψ =0. After nucleation occurs, the Meissner phase is restored, with a finite ψ value (in fact, basically with its zero field value).

2.2.4 Surface nucleation

When the magnetic field is applied parallel to the surface of a superconductor bounded by a dielectric, nucleation is modified by the boundary condition $\nabla_{\perp}\psi = 0$. Nucleation occurs at the field $H_{c3} = 1.69H_n$, in the form of a tegument of thickness ξ. If the superconductor is Type II ($H_n > H_c$), this tegument is stable between H_{c3} and H_{c2}, where the bulk vortex phase nucleates. If $(H_c/1.69) < H_n < H_c$ (this case is sometimes called Type I$\frac{1}{2}$ superconductor), the tegument is stable between H_{c3} and H_c where the Meissner phase nucleates. If $H_n < (H_c/1.69)$, nucleation of the surface tegument is immediately followed by nucleation of the Meissner phase.

2.3 Experimental results for the coherence length

2.3.1 Boundary effect experiments

Surface nucleation has been studied in Pb thin films deposited onto glass cylinders, by measuring the resonance frequency of a circuit composed of a coil wound around the cylinder in parallel with a capacitance. The outer surface of the Pb film was coated with a pair breaking normal metal. The imposed gradient on the order parameter quenches surface nucleation on that side of the film (Fig. 2.4).

When a magnetic field applied parallel to the axis of the cylinder exceeds H_c, it remains screened only by the inner tegument. The change in the self-inductance of the coil produces a change in the resonant frequency of the LC circuit. This change $\Delta\nu$ varies with the thickness of the film. By repeating the experiment for different film thicknesses t, and plotting $\Delta\nu(t)$, one observes a linear variation that extrapolates back to $\Delta\nu$=0 at a finite thickness, which is that of the tegument. This experiment provides a direct measurement of the coherence length (Fig. 2.5).

2.3.2 Nucleation field measurements

Experimental methods for the determination of ξ from nucleation field measurements are different for Type II and Type I superconductors.

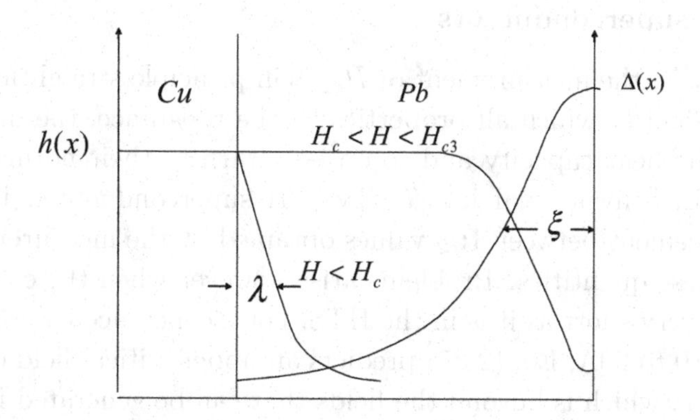

Figure 2.4: An over-coating of Cu quenches surface superconductivity on the outer side of a Pb film deposited on a dielectric cylinder. At $H = H_c$ the field penetrates up to the inner surface tegument. The change in inductance is detected by an LC circuit.

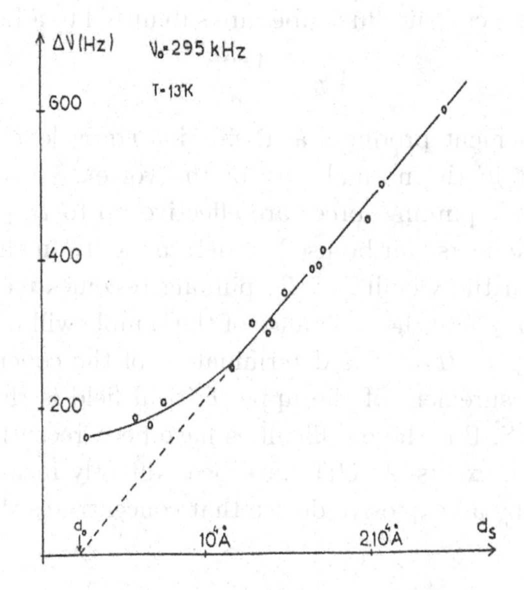

Figure 2.5: The change in the resonant frequency of the LC circuit at H_c measured as a function of the Pb film thickness extrapolates back to a finite value which is of the order of the coherence length.

Type II superconductors

In Type II's, the measurement of H_{c2} is in principle straightforward. It is the field at which all properties — the resistance, the magnetization, the heat capacity and so on — return to their normal state values. In conventional low T_c Type II superconductors, there is good agreement between H_{c2} values obtained by the measurement of any of these quantities. Problems arise however when the coherence length is very short as it is in the HTS. For a coherence length of the order of 10 to 20Å, Eq. (2.27) predicts an upper critical field of more than 100T, which is beyond the fields that can be generated in most laboratories. Measurements must then be carried out in the vicinity of the critical temperature, where the critical field is accessible. But fluctuation effects (see next chapter) then smear out the transition. In addition, at fields of order H_{c2}, the magnetization transition to the normal state is difficult to pinpoint. One has then to resort to resistive measurements, which are not an equilibrium property. In the presence of a current, flux tubes are submitted to a Lorentz force:

$$\mathbf{F}_L = \frac{\mathbf{J}\mathbf{x}\Phi_0}{c}$$

and their movement produces a dissipation equivalent to that due to the current in the normal core of the vortex. In conventional superconductors, pinning forces are effective up to H_{c2}, so that resistive measurements can be used to determine it. In the HTS, and in particular in the vicinity of T_c, pinning becomes inefficient at a field $H_{irr} \ll H_{c2}$, and the resistance of the sample will not in general remain zero up to H_{c2}. The determination of the coherence length through a measurement of the upper critical field is thus problematic in the HTS. But these difficulties have been recently overcome, critical fields in excess of 100T have been directly measured at low temperatures by an explosive device that concentrates the field for a short time.

Type I superconductors

In Type I superconductors, where the normal state can persist below H_c in a metastable state, nucleation of the superconducting state will occur preferentially at defects before the theoretical nucleation field will be reached. It will then propagate in the rest of the sample, giving a small supercooling effect, and a wrong value for ξ. This problem can be avoided by using samples consisting of small grains disconnected from each other: a defect in one grain will trigger the transition of that grain, but not of the rest of the sample. In addition, when $\kappa < 1.69/\sqrt{2}$, the measured nucleation field is always equal to H_{c3}, because the superconducting state will immediately propagate in the bulk after it has nucleated at the surface, at the equator of the grain. This correction is taken into account when calculating the value of ξ.

2.3.3 Comparison with theoretical predictions

We now have the tools to examine whether the inverse proportionality between the gap and the coherence length predicted by Eq. (2.1) holds experimentally. For a BCS condensation $2\Delta=3.5k_BT_c$. Since the Fermi velocity is close to $1\cdot10^8$cm/sec for most metals, Eq. (2.1) can then be rewritten as:

$$\xi \cdot T_c \approx 1.3\mu\text{m} \cdot K \tag{2.32}$$

To examine whether this coherence law holds, we have collected in Table 2.1 experimental values of the $\xi{\cdot}T_c$ product for a wide variety of superconductors. For low T_c metals, it does in fact converge remarkably well towards the value predicted by the above relation. This provides a direct and fundamental verification that condensation in these materials is indeed of the BCS type, pairs forming and condensing at the same temperature.

However, going down the table in order of increasing T_c values, we note a general trend towards decreasing values of $\xi \cdot T_c$. Several factors can explain this trend:

(i) due to strong coupling effects, the gap increases somewhat faster than T_c, an effect that we have neglected to take into account.
(ii) the electron-phonon interaction renormalizes the value of ξ by a factor $z=(1+\lambda_{\text{ep}})^{-1}$, where λ_{ep} is the electron-phonon interaction parameter.

Table 2.1: The BCS law predicting that the product of the coherence length by the critical temperature should be close to $1\mu\text{m}\cdot T_c(\text{K})$ is well verified for low T_c superconductors in the clean limit. It is not verified in high critical temperature superconductors. After G. Deutscher, in "Concise Encyclopedia of magnetic and superconducting materials", Ed. Jan Evetts, Pergamon Press.

	T_c (K)	ξ (μ m)	T_c (K) $\cdot\xi(\mu$ m)
Aluminum	1.19	1.2	1.4
Indium	3.40	0.33	1.1
Tin	3.72	0.26	0.97
Gallium	5.90	0.16	0.94
Lead	7.20	0.080	0.58
Niobium	9.25	0.035	0.32
$PbMoS_8$	15	0.0025	0.04
Nb_3Sn	17	0.0040	0.07
$C_{60}Rb_3$	31	0.0023	0.07
YBa_2Cu3O_7	93	0.0015	0.14
$Bi_2Sr_2CaCu_2O_8$	93	0.0010	0.09

These two corrections explain satisfactorily the decrease of $\xi \cdot T_c$ down to Pb. In addition:

(iii) in transition metals such as Nb, band structure effects significantly reduce the value of the Fermi velocity below the free electron value that we have used.

It is however doubtful whether these corrections are sufficient to explain the much lower values of the product $\xi \cdot T_c$ characteristic of the organic and High T_c superconductors. For T_c values higher than 20 to 30K, the coherence length appears not to change much. It is therefore legitimate to consider the possibility that condensation in these materials may not be of the BCS type.

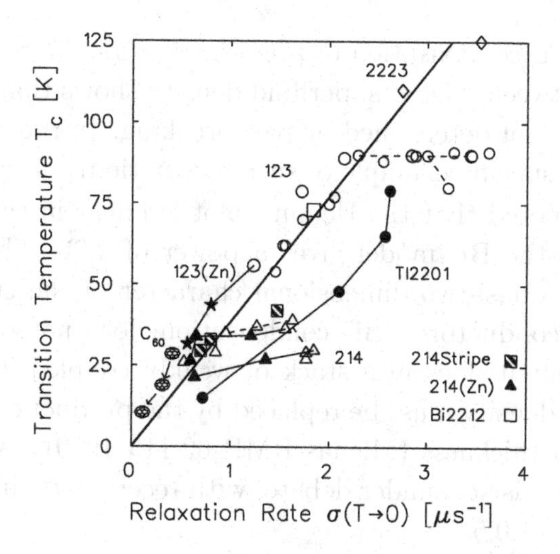

Figure 2.6: Transition temperature of several cuprates as a function of the muon spin relaxation rate, itself proportional to λ^{-2}. The dependence is linear in underdoped cuprates. Note the special behavior of $YBa_2Cu_3O_{7-\delta}$ for which T_c saturates for overdoped samples (δ <0.07) as the superfluid density keeps increasing. After Y. Uemura, Solid State Commun. **126**, 23 (2003).

2.4 Penetration depth and critical temperature

2.4.1 The Uemura law

As first noted by Uemura, there exists for the High T_c cuprates and other "exotic" superconductors, a remarkable correlation between the critical temperature and the superfluid density, as determined from a measurement of the penetration depth, Eq. (1.33).

For oxide superconductors, it was found that (Fig. 2.6):

$$T_c \propto \lambda(0)^{-2} \propto \frac{n_s}{m^*}$$

This empirical law reminds us of Eq. (1.13) for the Bose–Einstein condensation temperature, in the sense that it is a power law, if we identify the superfluid density with the boson gas density. The Uemura law, together with the violation of the BCS law of proportionality between ξ^{-1} and T_c, is an indication for a trend towards a Bose–Einstein condensation of preformed pairs. More generally, a correlation between T_c and superfluid density shows that the critical temperature is not determined by pair breaking. In the next section, we examine a specific example of such a situation.

It will be noted that the Uemura plot is linear in the superfluid density, while the BE model gives a power of 2/3. This has been related to the quasi-two dimensional character of the cuprates and organic superconductors. BE condensation does not occur in two dimensions, but it does in a stack of weakly coupled 2D slabs. In that case, the density must be replaced by the product of the density times the slab thickness (Micnas RMP 62,113 1990). Actually, the exact power law is still under debate, with recent experiments giving a power index of 0.5.

2.4.2 Granular superconductors

We consider an assembly of grains of a superconducting material having a bulk critical temperature T_{c0}, weakly coupled together. The size of each grain is smaller than the coherence length of the parent material, so that the amplitude and the phase of the order parameter must be constant within each grain. In the spirit of the GL theory, we present for the free energy:

$$F_s = F_n + A\sum_i |\Delta_i|^2 + \frac{B}{2}\sum_i |\Delta_i|^4 + C\sum_{i,j} |\Delta_i - \Delta_j|^2 \tag{2.33}$$

where the index i runs over all grains, and the indices i, j over nearest neighbor grains. This expression is written in terms of the pair potential Δ rather than of the wave function of the condensate. The amplitude of the pair potential has the dimension of an energy. It is identical to the gap for homogeneous superconductors. We consider the limit where the grains are large enough so that the amplitude of Δ has its equilibrium value Δ_0 in each grain. In the limit where $C \to 0$, the grains are simply disconnected. The amplitude of Δ is the same on each grain, but the phases are random. When C is small, in the sense that:

$$C \ll |A| \tag{2.34}$$

the last term on the r.h.s. of Eq. (2.33) represents a small but finite free energy increase due to phase fluctuations from grain to grain. We can then write:

$$C \sum_{ij} |\Delta_i - \Delta_j|^2 = 2C|\Delta|^2 \sum_{ij} \cos(\theta_i - \theta_j) \tag{2.35}$$

where θ_i is the phase of the pair potential on grain i. If the phases are completely random, this term is of order $2C|\Delta|^2$. It introduces a small shift in the temperature below which there is a gain in free energy for a finite value of Δ. This temperature T^* is now given by the condition:

$$A(T^* - T) + 2C = 0 \tag{2.36}$$

At T^*, superconductivity appears in the grains, but the phases are still unlocked. At sufficiently low temperature, when $k_B T$ per grain is of order $C|\Delta|^2$, it will become more advantageous to lock the phases of all grains: the system will then be in the coherent superconducting state.

The interesting point is that this granular system has not one, but two characteristic temperatures:

(i) a temperature $T^* < T_{c0}$, below which intra-grain superconductivity appears.
(ii) a temperature T_c, of order $C|\Delta|^2$, below which inter-grain superconductivity is achieved. According to the above argument, this temperature should be proportional to C.

One can generalize Eq. (2.33) to the case where there is a finite vector potential $\mathbf{A}$. The inter-grain term then becomes:

$$H_{ij} = 2C|\Delta|^2 \sum_{ij} \cos(\theta_i - \theta_j - A_{ij}) \tag{2.37}$$

where:

$$A_{ij} = \frac{2e}{hc} \int_i^j \mathbf{A} \cdot d\mathbf{l} \tag{2.38}$$

From Eqs. (1.30), (1.33) and (2.14), the penetration depth is given in terms of the coefficient C in (2.31) by:

$$\lambda^{-2} = \frac{64\pi e^2}{\hbar^2 c^2} C|\Delta|^2 \tag{2.39}$$

Hence, in our granular superconductor:, $\lambda^{-2} \propto T_c$ Chakraverty and Ramakrishnan give:

$$k_B T_c = \frac{\Phi_0^2}{15\pi^3} \frac{\xi_0}{\lambda^2} \tag{2.40}$$

The remarkable property of the granular system is that the penetration depth is directly related to the coherence critical temperature T_c, similarly to what was observed by Uemura for a number of non-conventional superconductors that do not follow the BCS predicted correlation between T_c and the coherence length. It is one specific example of a superfluid whose condensation temperature is limited by a small superfluid density rather than by the value of the energy gap. Equation (2.40) is in good agreement with data on granular Aluminum, as we shall see in Chapter 4 where we review some properties of granular superconductors.

2.5 Further reading

For a complete description of the condensed state and of Ginzburg Landau equations, see P.G. de Gennes, "Superconductivity of Metals and Alloys", Chapter 4 and Chapter 6 (W.A. Benjamin, Inc., New York 1966).

For a review on nucleation fields, see J.P. Burger and D. Saint-James, in "Superconductivity", Ed. R.D. Parks, M. Dekker, Inc., New York (1969), p. 977.

For a review on coherence length effects, see G. Deutscher in "Encyclopedia on Superconductivity and Magnetism", Ed. Jan Evetts, Pergamon Press.

For discussions of the effect of a reduced superfluid density on the superconducting transition, see B.K. Chakravery and T.V. Ramakrishnan, Physica C **282-287**, 290 (1997); Y. Uemura, Solid State Commun.**126**, 23 (2003) and references therein, V. Emery and S. Kivelson, Nature **374**, 434 (1995).

Chapter 3

The phase transition

The transition to the superconducting state belongs to the general class of second order phase transitions. The order parameter involved in this transition has two components: an amplitude and a phase. Models for such phase transitions in three dimensions are known as 3D XY models. They describe for instance the divergence of the heat capacity of liquid Helium when it undergoes the transition to superfluidity. This divergence reflects the existence of large thermodynamical fluctuations of the order parameter, a general property of second order phase transitions near the critical point.

As we have seen in the preceding chapter, fluctuation effects are neglected in the GL theory of superconductivity. In this chapter, we shall clarify in which case this is justified, and when it is not. But before we deal with the issue of the importance of fluctuations in the superconducting phase transition, we shall examine how one can determine experimentally the fundamental properties of this transition, namely the loss of entropy and the gain in free energy.

3.1 Free energies

As in all phase transitions, there is a loss of entropy ΔS and a gain in free energy ΔF between the high temperature normal state and the low temperature superconducting phase. ΔS and ΔF are the primary characteristics of the transition. Here, we are only considering the electronic component, assuming that the ions dynamics is basically not affected by this transition.

From the relations:

$$S = -\frac{dF}{dT} \tag{3.1}$$

and:

$$C = T\frac{dS}{dT} \tag{3.2}$$

We can, starting from a measurement of $C(T)$ of the electrons, calculate the temperature dependence of $S(T)$ and $F(T)$:

$$S(T) = \int_0^T \left(\frac{C}{T'}\right) dT' \tag{3.3}$$

$$F(T) = -\int_0^T S(T')dT' \tag{3.4}$$

The transition being of the second order, there is no latent heat and no discontinuity of the entropy at the transition.

What is of interest to us are the differences $\Delta C(T)$, $\Delta S(T)$ and $\Delta F(T)$ between the superconducting and the normal states, if the latter would be made to persist below T_c, for instance by the application of a strong magnetic field. In usual metals, we can also use the quasi-particle approximation:

$$C_N(T) = \gamma T \tag{3.5}$$

at all temperatures that are well below the Fermi temperature. To obtain $C_N(T)$ below T_c , we just extrapolate the linear high temperature behavior measured above T_c (Fig. 3.1). If the quasi-particle approximation is not valid, we need to find a way to measure $C_N(T)$ below T_c to obtain $\Delta S(T)$ and $\Delta F(T)$.

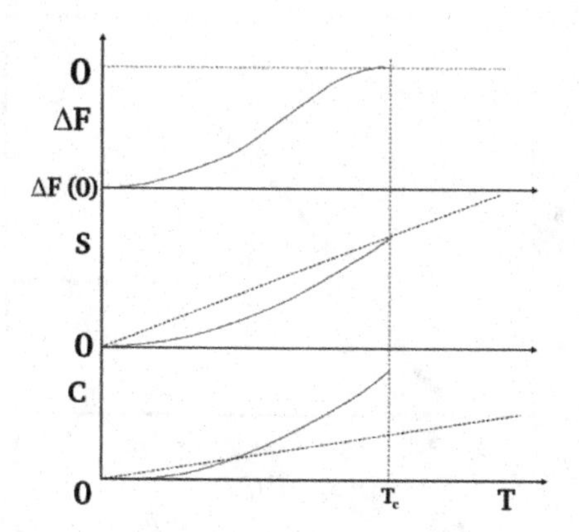

Figure 3.1: Electronic heat capacity C, entropy S, and difference between the superconducting and normal state free energies ΔF as a function of temperature. Dashed lines are for the normal state. $\Delta F(0)$ is the free energy gained in the superconducting state at zero temperature.

3.1.1 Metals: mean field behavior

The experimentally observed behavior of the electronic heat capacity in metals is characterized by a linear variation above T_c, a finite jump at T_c, a finite slope just below T_c, and an exponential decrease at low temperatures, as shown in Fig. 3.1. We can then construct $\Delta S(T)$ and $\Delta F(T)$.

The result is in line with the predictions of the GL theory. From:

$$\Delta F = -\frac{a^2}{2b} \tag{3.6}$$

we get:

$$\Delta F(T) = -\frac{\bar{a}^2}{2b}(T - T_c)^2 \tag{3.7}$$

$$\Delta S(T) = \frac{\bar{a}^2}{b}(T - T_c) \tag{3.8}$$

$$\Delta C(T \leq T_c) = \frac{\bar{a}^2}{b}T \tag{3.9}$$

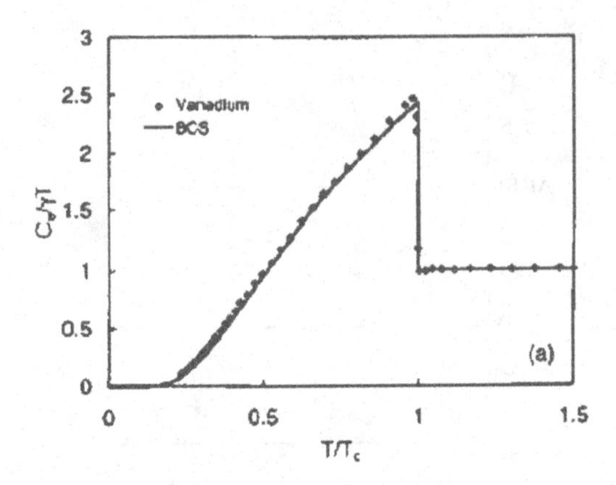

Figure 3.2: In a conventional metal-superconductor such as Vanadium, the fit between experimental heat capacity data and mean field theory is excellent. Note the absence of any measurable fluctuation effects above T_c. After Junod *et al.*, Physica C **317–318**, 333 (1999).

The GL theory predicts below T_c a parabolic behavior of ΔF, a linear behavior of ΔS, and at T_c a finite jump of ΔC (Fig. 3.1). These predictions very well fit the experimental results (Fig. 3.2). Since the GL and BCS theories neglect all fluctuations of the superfluid density around its mean value (mean field approximation), we conclude from the excellent fit that this approximation is sufficient to describe the transition in metals.

3.1.2 Examples of non-mean field behavior

We now show two examples of non-mean field behavior that do not fit the above description.

Figure 3.3 shows the heat capacity of small grains of Aluminum, embedded in an insulating matrix (here Germanium). Instead of a jump, there is a broad transition. As we shall see in details in the next section, this broadening is due to the finite size of the grains. It makes it possible for the amplitude of the pair potential to fluctuate around its equilibrium value in the condensed state, and to have a finite mean square value in the normal state. In addition, we also note that the transition starts well above the critical temperature of

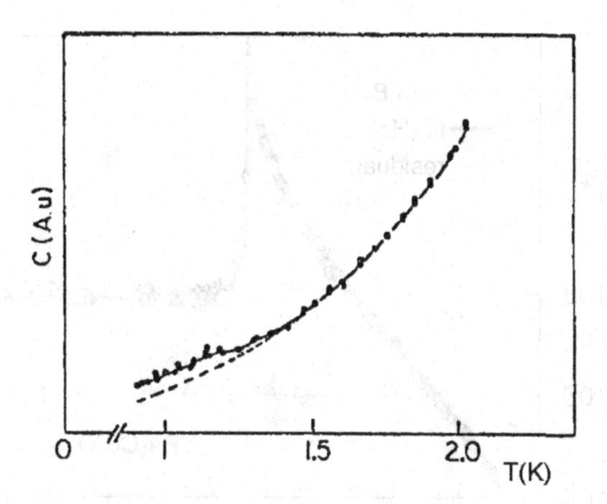

Figure 3.3: Heat capacity transition in a thin film of granular Al composed of uncoupled grains of about 10 nm in diameter. This film is insulating and has no resistive transition. Most of the heat capacity is that of the substrate. After Y. Shapira, PhD Thesis, Tel Aviv 1982.

bulk Aluminum (1.2K). Possible reasons for this enhancement will be discussed in Chapter 8 where we deal with interaction mechanisms.

A different deviation from mean field behavior is seen in the transition of high T_c cuprates, as shown in Fig. 3.4. In addition to some broadening observed above T_c, the shape of the main transition is modified. Instead of a jump, it looks more as a narrow peak. Qualitatively, it reminds of the specific heat peak seen at the transition of superfluid Helium. Again, we are dealing with a fluctuation effect, but now in a three-dimensional medium. Fluctuations in 3D lead to a divergence of the heat capacity. As we shall see in the next section, this divergence, expected from the 3D XY model, is visible in the high T_c cuprates because of their short coherence length.

3.1.3 The anomalous normal state in the ceramics

Another kind of deviation from the simple behavior shown Fig. 3.1, now affecting the normal state, occurs in certain superconducting ceramics. Unlike usual metals and alloys, whose properties are very

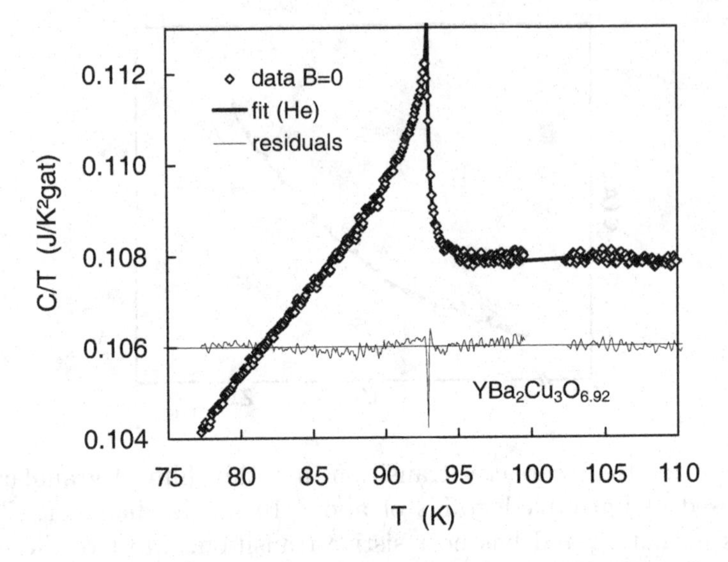

Figure 3.4: Heat capacity of an optimally doped $YBa_2Cu_3O_{7-\delta}$ sample. Note the sharp peak and the fit to the transition of superfluid Helium. After Junod *et al.*, Physica C **280**, 214 (2000).

resilient to the introduction of a small amount of impurities or a small change in the concentration of the constituents, the properties of the ceramics may vary drastically upon a small change of chemical composition, for instance oxygen content.

We see an example of this sensitivity when we compare the heat capacity of YBCO samples depleted in oxygen concentration (Fig. 3.5), to that of the sample with the concentration 6.95 shown in Fig. 3.4. In addition to the lowering of T_c and the reduction of the heat capacity peak, there is also a change in the high temperature behavior. The entropy $S(T)$ is linear in optimally doped and overdoped samples (Fig. 3.5(a)), but not for the depleted ones (Fig. 3.5(b)). The conclusion reached by Loram *et al.* from their data is that in oxygen depleted samples thc normal state, meaning the state above T_c, is not the ordinary metallic one. Another point concerns the transition itself. Compared to that of the optimally doped and overdoped samples, the peak at T_c in underdoped

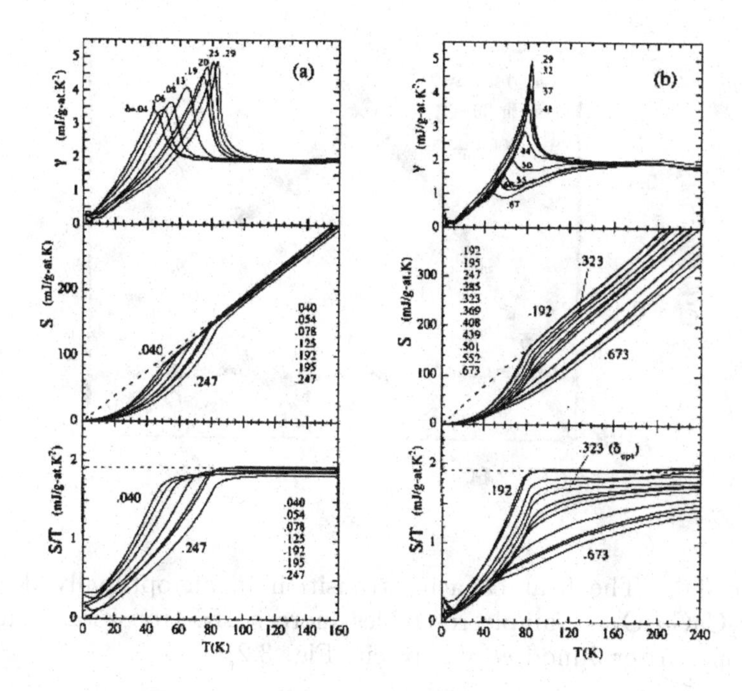

Figure 3.5: Heat capacity transitions in a series of $Y_{0.8}Ca_{0.2}Ba_2Cu_3O_{7-\delta}$ samples. Strong overdoping is achieved with Ca substitution. Note the conventional transition in overdoped samples (a), the sharp peak at optimum doping (maximum T_c), the small peak as well as the anomalous normal state behavior in underdoped samples (b). After Loram *et al.*, J. Phys. Chem. Solids **59**, 2091 (1998).

samples is quite reduced. The loss of entropy is smaller. A tempting interpretation would be that the loss of entropy in the electron gas is taking place in two steps. First, pairs form at high temperatures, and then condense at T_c. The entropy loss at the transition is then smaller than in the BCS condensation, where pairs form and condense simultaneously. This is a scenario that we have already considered. Another possibility, proposed by Loram *et al.*, is that a gap has opened up in the normal state, so that only a fraction of the carriers takes part in the transition. These questions are in the forefront of high T_c research these days. We shall return to them in later chapters.

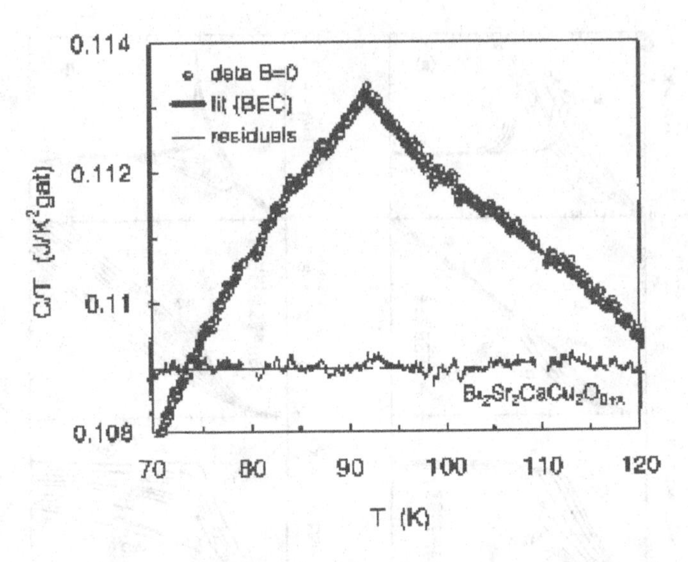

Figure 3.6: The heat capacity transition in an optimally doped $Bi_2Sr_2CaCu_2O_{8+\delta}$ sample resembles more a BE transition than a BCS one. After Junod *et al.*, op. cit. Fig. 3.2.

An even more radically different form of heat capacity transition is observed in $Bi_2Sr_2CaCu_2O_{8+\delta}$. There is no more heat capacity jump at the transition, but rather a cusp. The transition is better fitted by a BE condensation than by a BCS one (Fig. 3.6).

3.2 Fluctuations

3.2.1 The small grain case

Let us now consider a small grain of a metal-superconductor ($\xi \gg a$), of size $d<\xi$. We might worry about the boundary condition that applies to the superfluid density at the surface of the grain. But the detrimental effect of the surface is felt over the atomic scale a, and we can guess that it will be averaged over the scale of ξ. So, if we consider a grain of a low T_c superconductor, with a coherence length orders of magnitude larger than the interatomic distance, as we do here, the effect should be small. We shall assume in the following that the superfluid density in the grain is uniform.

Now, let us assume that the value of the order parameter deviates momentarily from this equilibrium value;

$$\Psi = \Psi_0 + \delta\Psi \tag{3.10}$$

To be specific, let us consider the case $T > T_c$, where $\Psi_0 = 0$, and calculate the cost in free energy of this fluctuation. If its amplitude is small, we can neglect the fourth order term in the GL expansion, and obtain:

$$\delta F = a \langle \mid \delta\Psi \mid^2 \rangle d^3 \tag{3.11}$$

The grain being near the bulk critical temperature, this increment of the free energy is provided by the thermal energy $k_B T_c$:

$$\langle |\delta\Psi|^2 \rangle \propto \frac{k_B T_c}{a d^3} \tag{3.12}$$

We can easily repeat this calculation below T_c and obtain:

$$\langle |\delta\Psi|^2 \rangle \propto \frac{k_B T_c}{2|a| d^3} \tag{3.13}$$

The amplitude of the fluctuations (squared) diverges as $(T - T_c)^{-1}$. We must check under what conditions $|\delta\Psi| < \Psi_0$, which was our initial assumption:

$$\frac{k_B T_c}{d^3} < \frac{a^2}{b} \tag{3.14}$$

This corresponds to a temperature range δT such that:

$$\left(\frac{\delta T}{T_c}\right) > \left(\frac{k_B T_c}{\Delta F(0) d^3}\right)^{1/2} \tag{3.15}$$

where $\Delta F(0) = (\bar{a}^2/b)$ is (twice) the difference in free energy between the superconducting and the normal states (the condensation energy, as defined earlier) per unit volume, extrapolated down to T=0.

As we have seen in Chapter 1, this condensation energy is of the order of $\left(\frac{\Delta^2}{E_F}\right)$ per pair. Using the free electron expression for the normal state density of states, $N(0) = \frac{3}{2}\frac{n}{E_F}$ where n is the electron density, and recalling that $k_B T_c$ is itself of the order of Δ, we see that the denominator of the r.h.s. of Eq. (3.15) is essentially the number of pairs in the grain at T=0. If this number is much larger than unity, the mean field approximation will be valid in most of the temperature range, except very near T_c. But if this number approaches unity, fluctuations of the order parameter will be so large that there will be no range of temperature where our approximation holds. In fact, it is sort of obvious that if there is on the average less than one pair per grain, superconductivity cannot survive. Another way to express this result is to say that our treatment will have a range of validity, as long as the number of free electrons in the grain will be larger than $(\frac{E_F}{\Delta})$. Yet another way to put it, is to say that the spacing between the electronic levels in the grain, $(N(0)d^3)^{-1}$ must be smaller than Δ.

We have obtained an interesting result: if we make the grain small enough, some sizeable short range order appears *above* T_c: the transition can no longer be described as a mean field one.

The effect of thermodynamical fluctuations can be observed in the heat capacity transition. When the condition (3.15) holds, we calculate it from the approximate partition function:

$$Q = \int_0^\infty d \mid \delta\Psi \mid^2 \exp - \left(\frac{a \mid \delta\Psi \mid^2 d^3}{k_B T} \right) \tag{3.16}$$

and by taking derivatives from the free energy:

$$F = k_B T \ln Q \tag{3.17}$$

one then obtains:

$$\delta C = \Delta C_{MF} \left(\frac{\epsilon_c}{\epsilon} \right)^2 \tag{3.18}$$

where ϵ is the reduced temperature scale:

$$\epsilon = \frac{T - T_c}{T_c} \tag{3.19}$$

and ϵ_c is given by Eq. (3.15) when one replaces the inequality sign by an equality.

In fact, for the case of the small grain the heat capacity transition can be calculated exactly in the entire temperature range across T_c. It goes from a sharp jump for a large grain, to a washed out transition when the grain approaches the critical size where there is one pair per grain. This compares well with the heat capacity transition measured in small Al grains.

One could have naively thought that superconductivity would be quenched in a grain smaller than the coherence length. In fact, this is not the case here, where we are considering a grain made of a superconductor having a large coherence length — simply because in that case, as we have seen earlier, there are so many pairs within a coherence volume. In a grain smaller than the coherence length, there are still enough pairs around to produce the superconducting transition.

Thermodynamic fluctuations of the order parameter — if nothing else — will quench superconductivity if the grain is made small enough. If we consider for example the case of Al, where Δ is about four orders of magnitude smaller than the Fermi energy, this will happen when there are around $1 \cdot 10^4$ free electrons in the grain. The corresponding grain size is of about 50Å. Al samples consisting of grains of about that size, electrically uncoupled, can actually be made. For the sample shown in Fig. 3.3, the grain size is about twice that size: the transition is already broadened, but still quite visible. The study of fluctuation effects is one of the reasons for studying granular superconductors.

3.2.2 Three-dimensional fluctuations: quasi-mean field treatment

In a three-dimensional sample, fluctuations of the order parameter will in general vary in space. The GL free energy form can still be used, if we assume that fluctuations are small in amplitude and uncorrelated, which will be the case if we are not too close to T_c

(quasi-mean field approximation). We then write:

$$\Psi = \Psi_0 + \delta\Psi(\mathbf{r}) \tag{3.20}$$

Fluctuations occur both above and below T_c. Above T_c, Ψ_0=0, and within the approximations made we can write:

$$\Delta F = a \mid \Psi \mid^2 + c \mid \nabla\Psi \mid^2 \tag{3.21}$$

Here we have taken into account only amplitude fluctuations, again within the assumption that fluctuations are few and far apart, therefore uncorrelated.

The spirit of this approximation is the same as the one we have made in the study of the small grain. But now, the spatial dependence of the fluctuation amplitude cannot be neglected, and in fact plays a major role. The most likely fluctuations to occur are those for which, for a given value of the fluctuation amplitude, the two terms of the r.h.s. of Eq. (3.21) will be of the same order of magnitude. This means that they occur over a length scale ξ (the coherence length) such that:

$$\xi(T)^2 = \frac{c}{\mid a \mid} \tag{3.22}$$

The free energy cost of a typical fluctuation should be approximately its free energy per unit volume, times its volume:

$$\delta F \cong a \mid \delta\Psi \mid^2 \xi^3(T) \tag{3.23}$$

This quantity should be of the order of the available thermal energy, $k_B T_c$. Thus:

$$\langle \mid \delta\Psi \mid^2 \rangle \cong \frac{k_B T_c}{a\xi(T)^3} \tag{3.24}$$

Finally, the free energy per unit volume is:

$$\Delta F \cong \frac{k_B T_c}{\xi(T)^3} \tag{3.25}$$

The partition function can again be calculated. With the reduced temperature scale introduced before, and writing:

$$\xi(T) = \xi_0 \epsilon^{-\frac{1}{2}} \tag{3.26}$$

the result for the heat capacity above T_c is:

$$\Delta C = \Delta C_{MF} \left(\frac{\epsilon_c}{\epsilon}\right)^{\frac{1}{2}} \tag{3.27}$$

where:

$$\epsilon_c = \eta \left(\frac{k}{\xi_0^3 \Delta C_{MF}}\right)^2 \tag{3.28}$$

where $\eta = (32\pi^2)^{-1}$. Since we have assumed that fluctuation effects are small, this result is valid when $\epsilon \gg \epsilon_c$, outside of what is called the *critical region*. Again within the framework of the small fluctuation approximation, this result is only applicable if $\epsilon_c \ll 1$. Below T_c, a similar expression is found, but with an amplitude that is different by a factor of $\sqrt{2}$.

The quasi mean field approximation is a useful tool to estimate the importance of fluctuations in a given superconductor, once the values for the coherence length and of the heat capacity jump are known. We notice particularly the strong dependence of ϵ_c on the coherence length. The critical region is small for long coherence length, low T_c superconductors, a typical width being of the order of 10^{-10}, much too small to be accessible experimentally. In that case, mean field theory is sufficient to describe thermodynamical properties. But this is not the case in high temperature superconductors.

3.2.3 The heat capacity transition in the High T_c: the case of YBCO

A close look at the heat capacity transition in optimally doped YBCO samples (Fig. 3.4) shows that it starts several degrees above T_c, and presents a rather sharp peak, with an increasing slope (dC/dT) as T_c is approached. These are strong indications that thermodynamical fluctuations are playing an important role in the transition. We can use Eq. (3.28) to estimate the width of the critical region in a typical cuprate. Using the GL expression for ΔC and extracting the value of $\left(\frac{a^2}{b}\right)$ from the condensation energy per unit volume itself given by $\left(\frac{H_c^2}{8\pi}\right)$ with H_c=1T, and with $\xi_0^3 = \xi_{ab}^2\xi_c$, ξ_{ab} =10Å, ξ_c=2Å, we obtain $\epsilon_c \simeq 1\cdot 10^{-2}$. This is only a rough estimate, but for a typical cuprate having a critical temperature of 100K, the width of the critical region is now predicted to be of the order of 1K, instead of 10^{-10}K in a low temperature superconductor. Quasi-mean field theory, Eq. (3.27), predicts that at 10K above T_c there should be an increase of the electronic heat capacity of about 30% of ΔC, as compared to the normal state value. This is larger than observed experimentally, but there is a visible heat capacity tail up to several degrees above T_c, Fig. 3.4.

We have of course no reason to expect that quasi-mean field theory can describe the entire transition, since its validity is limited to the range $\left(\frac{\epsilon}{\epsilon_c}\right) \gg 1$. If we wish to describe the data close to T_c, we need a model that describes large (critical) fluctuations. A general expression describing them is of the form:

$$\frac{C^{\pm}}{T} = \frac{1}{T}C_{\text{background}} + \frac{A^{\pm}}{\alpha} \mid \epsilon \mid^{-\alpha} \tag{3.29}$$

$C_{\text{background}}$ is obtained from a measurement of the heat capacity in the presence of a very strong field, sufficient to completely quench superconductivity in the vicinity of T_c. The + and - superscripts correspond respectively to temperatures above and below the transition: In the 3D XY model, which describes a second order phase transition in three dimensions with a two component order parameter (here the amplitude and the phase of the order parameter), $\frac{A^+}{A^-}$=1.054±0.001 and α =-0.01285±0.00038. The temperature dependence of the coherence length is determined by a critical exponent ν different from the value 1/2 of the mean field theory. It is related to the exponent α and to the dimensionality D by the hyperscaling relation, $\alpha + D\nu = 2$,which gives in 3D, ν=0.67.

The mean field jump can be extracted from Eq. (3.29). In the limit $\mid \alpha \ln \mid \epsilon \mid\mid \ll 1$, it can be rewritten as:

$$\frac{C^-}{T} \cong \frac{1}{T} C_{\text{background}} + A^+(J - \ln \mid \epsilon \mid), \; T < T_c \tag{3.30}$$

$$\frac{C^+}{T} \cong \frac{1}{T} C_{\text{background}} + A^+(- \ln \mid \epsilon \mid), \; T < T_c \tag{3.31}$$

where $J = (A^-/A^+ - 1)/\alpha \cong 4$ in the 3D XY model. The mean field jump is the difference between values of C/T measured below and above T_c for the same absolute value of the reduced temperature scale. From the above expression, it is equal to A^+J.

Junod *et al.* have fitted their data to these expressions, letting J be a free parameter. For samples of YBCO at optimum doping, they found for J a value very close to 4, as predicted by the 3D XY model. However, this is the only case where the model fits the data. For overdoped YBCO, obtained J values are larger than 4, and for underdoped YBCO they are smaller. Values larger than 4 may still be compatible with the 3D XY model. They mean that the observed divergence of the heat capacity is smaller than predicted, which may just be the result of the critical region becoming too narrow to be accessible experimentally. Values smaller than 4, on the other hand, are *not* compatible with the model, because the mean field jump is a robust feature, rather insensitive to experimental limitations that may arise in the vicinity of the transition. The discrepancy increases gradually with underdoping, J reaching a value as low as 2.4 in a sample having a critical temperature of 53K. The value of the mean field jump is shown for YBCO (Fig. 3.7) as a function of oxygen doping together with T_c.

The conclusion that optimally doped and probably overdoped YBCO belong to the same class as BCS superconductors, the only quantitative difference being its much shorter coherence length as compared to LTS, is also borne out by the amplitude of the mean

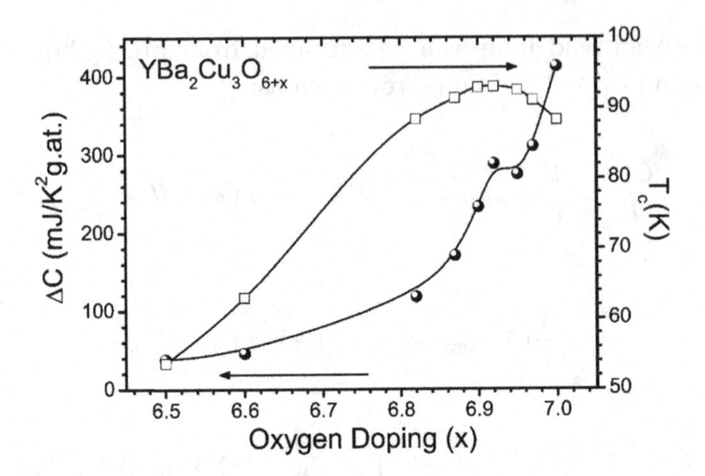

Figure 3.7: Mean field heat capacity jump ΔC and critical temperature of YBCO as a function of oxygen doping. Lines are guides to the eye. Note the sharp decrease of ΔC as one goes from fully oxygenated, overdoped, to slightly underdoped samples. Data taken from Junod *et al.*, see further reading for references.

field jump. GL theory tells us that $\Delta C/T_c$ is proportional to the condensation energy per unit volume, divided by the critical temperature squared. In a BCS superconductor, this ratio is a constant since the condensation energy is proportional to the square of the gap, itself proportional to the critical temperature. As remarked by Junod *et al.*, this ratio in overdoped YBCO is indeed the same as for Nb. At this point, we recall from the preceding chapter that the critical temperature of overdoped YBCO becomes independent from the superfluid density, as expected for a BCS superconductor.

In optimally doped and overdoped YBCO, the small coherence volume is thus the only reason for the observable effect of thermodynamical fluctuations on the transition. There may be only about 10 pairs in this volume, in sharp contrast with the case of the metal-superconductors, such as Al, where the number of pairs per coherence volume is of the order of 10^7! The small coherence volume in YBCO plays qualitatively the same role as the small grain size in the low T_c, in enhancing the effect of fluctuations.

On the basis of the same analysis, we calculate a condensation energy per particle for optimally doped YBCO on the order of 0.1 k_BT_c, and a condensation energy per coherence volume of the order of k_BT_c. Again, this is in sharp contrast to the case of a typical metal-superconductor such as Al, for which the condensation energy per particle is on the order of $1 \cdot 10^{-4} k_BT_c$, and the condensation energy per coherence volume many orders of magnitude larger than k_BT_c (in fact several orders of magnitude larger than the Fermi energy).

Underdoped YBCO samples do not belong to the class of BCS superconductors. Their behavior is closer to that of the highly anisotropic cuprates, described below.

3.2.4 The heat capacity transition in high anisotropy cuprates

In compounds such as $Bi_2Sr_2CaCu_2O_{8+\delta}$ (usually called Bi-2212), the heat capacity transition looks very much different from that in YBCO. It is nearly symmetrical above and below T_c (Fig. 3.6). This shape is typical for cuprates having a strong anisotropy between the in-plane and out-of-plane conductivities, coherence lengths ξ_{ab} and ξ_c, and penetration depth λ_{ab} and λ_c. In YBCO, the ratio (ξ_{ab}/ξ_c) is of about 5, while in Bi-2212 it is of about 30. If one nevertheless applies the same analysis to Bi-2212 as to YBCO, as Junod *et al.* have done, values of the parameter J smaller than one are obtained. Values obtained for the mean field jump are typically one order of magnitude smaller than for optimally doped to overdoped YBCO. At this point, it is worth recalling that this analysis assumes $\alpha \ll 1$. This assumption may not apply here. A fit to the more general expression Eq. (3.27) using the Bose–Einstein exponent $\alpha = -1$ is possible. Letting α be a free parameter lends a value of -0.7. The condensation is certainly not of the BCS type, not even in a border line way as it may be in underdoped YBCO. A fit to a Bose–Einstein condensation is better. Yet, similarly to what is seen in YBCO, the details of the transition are sensitive to the amount of doping. Fitted J values in overdoped Bi-2212 (0.8) are larger than in optimally doped samples (0.2). The trend towards a BE condensation is progressive as doping is decreased, just as is the deviation from BCS mean field is progressive in YBCO.

This difference in behavior is also apparent in the variation of the critical temperature with the superfluid density. In the Bi2212 compound, there is no range of superfluid density where it remains constant, contrary to what is seen in YBCO, see Fig. 2.6.

3.3 Condensation energies

In cuprates for which a mean field jump cannot be clearly identified (for optimally doped to underdoped Bi2212 or for strongly underdoped YBCO for instance), one cannot calculate the condensation energy from the GL theory. But this can still be done using heat capacity data in the entire temperature range, using the procedure described in Sec. 3.1.

A very complete set of such data has been given by Loram on a number of cuprates, at different doping levels. They have been obtained by a differential method, where the heat capacity of the background (phonons etc...) is subtracted by measuring simultaneously the heat capacity of "sister sample" in which superconductivity has been quenched by doping with a small concentration of Zn impurities.

Condensation energies in overdoped to underdoped YBCO have been obtained in this way by Loram *et al.* (see further reading for references) on the basis of the data shown in Fig. 3.5. The condensation energy per unit volume has a maximum for slightly overdoped samples, beyond which it varies with T_c as it does in a BCS superconductor. On the contrary, it decreases in the underdoped regime much faster than it would in a BCS superconductor (Fig. 3.8). One will notice that there is a close correlation between the doping dependence of the condensation energy and the value of the mean field heat capacity jump at T_c. Qualitatively, a similar trend is seen in the Bi2212 compound.

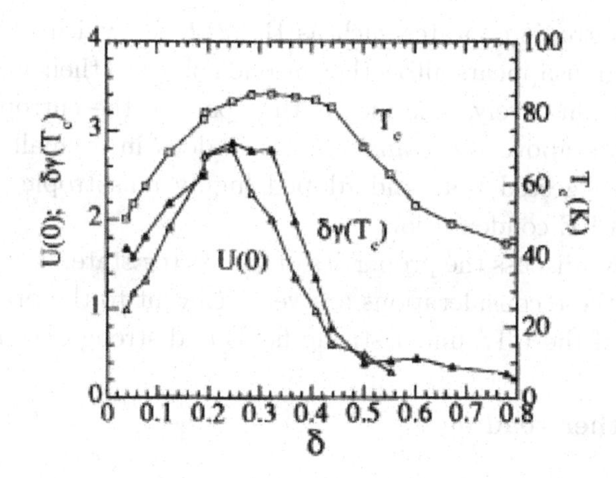

Figure 3.8: Critical temperature, heat capacity jump $\delta\gamma(T_c)$ (uncorrected for fluctuation effects) and condensation energy $U(0)$ obtained from the data of Fig. 3.5 on $Y_{0.8}Ca_{0.2}Cu_3O_{7-\delta}$ samples. After Loram *et al.*, Physica C **282–287**, 1405 (1997).

3.4 Summary

Thermodynamical fluctuation effects on the heat capacity transition are clearly observed in small superconducting grains and in the HTS. This is due to the small number of Cooper pairs in the relevant volume, that of the grain or that of the coherence volume.

There are substantial differences amongst cuprates. The transition in the less anisotropic YBCO, overdoped to optimally doped, falls into the same universality class as the BCS transition. It is as expected from the 3D XY model. The small coherence volume is sufficient to explain why thermodynamical fluctuations affect the heat capacity up to several degrees above the transition, in strong contrast with the LTS where the coherence volume is larger by many orders of magnitude and such effects are not observable. Deviations from this simple behavior are seen in underdoped YBCO. In addition to the small coherence volume effect, there is also a strong diminution of the heat capacity jump per unit volume as the doping level is reduced. This diminution is the dominant feature in the highly anisotropic cuprates such as Bi-2212, for which the heat capacity jump disappears altogether as one goes into their underdoped regime. Qualitatively, this means that part of the entropy loss already occurs above the transition, as it does in a small grain. It has been suggested that underdoped highly anisotropic HTS may approach a BE condensation.

When we discuss the properties of the vortex state, it will become clear that these considerations are very relevant to the practical applications of the HTS under strong fields and strong currents.

3.5 Further reading

The critical behavior of the heat capacity is described and discussed in great detail and depth by Junod *et al.* in Physica B **280**, 214 (2000). Data emphasizing the anomalous normal state behavior and its relation to superconductivity are given and interpreted in detail by Loram *et al.* in J. Phys. Chem. Solids **59**, 2091 (1998).

Chapter 4

Phase diagrams

For reasons that are still imperfectly understood, elevated critical temperatures are only observed in compounds that are close to a Metal to Insulator (M/I) transition. This is the case in materials as diverse as the cuprates, oxides such as KBaBiO, organic superconductors, the fullerenes and granular superconductors. Contrary to usual metals and alloys, the normal state properties of these materials, as well as their critical temperature, are very sensitive to their exact composition or other parameters that can control the distance from the M/I transition.

A simplified generic phase diagram of this class of superconductors is shown in Fig. 4.1. In the (T, x) plane, where x is a parameter that expresses the distance from the M/I transition, one can distinguish three phases. Superconductivity appears above a minimum value x_c. The critical temperature reaches a maximum at some value x_M and persists up to x_F. The normal state is metallic at $x > x_M$, and insulating at $x < x_M$. An antiferromagnetic state is often observed close to the M/I transition.

Insulators are of different kinds: band insulators, Anderson insulators where electron localization is due to disorder at the atomic scale, percolation insulators where disorder is on a macroscopic scale (namely much larger than atomic), and Mott insulators driven by electron-electron Coulomb repulsion. The kind of M/I transition we have in mind here is to that last kind of insulating state. Such a transition is expected when the carrier density is reduced to such

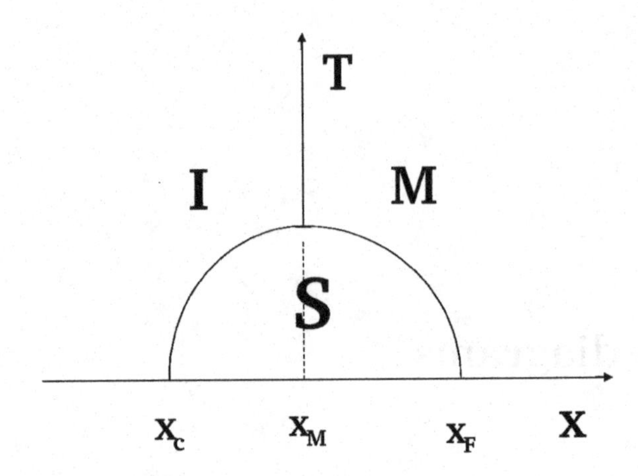

Figure 4.1: Schematic generic phase diagram of a superconductor near a Mott metal to insulator transition. Above the superconducting dome, the normal state is metallic above "optimum doping" and weakly insulating below it.

an extent that Thomas–Fermi screening becomes so weak that electrons start to feel strong on-site repulsion and become localized on individual atoms.

It is in general difficult, and in fact practically not possible, to start from an arbitrary metal and to bring it up to a M/I transition of that kind by the use of some control parameter. It is believed that a *gedanken* experiment by which one would start from a copper crystal and start to pull apart its atoms would result in such a transition. But we do not know how to perform such an experiment in the laboratory. An alternative way would be to mix atoms of the metal with "neutral" atoms, thus diluting the valence electrons. Such experiments have been performed, but at high dielectric volume fraction one would presumably completely transform the structure of the metal. For instance, the mixture may become amorphous.

In this chapter we give two examples of controlled Mott transitions: that of granular Aluminum, and that of a typical cuprate. In both cases superconductivity persists up to the transition, and in fact is enhanced rather than weakened in its vicinity (before being eventually quenched). In the granular case, the transition is triggered by electronic decoupling of small grains. Each of them is a "quantum dot", meaning that it is so small that the splitting between the electronic levels due to the finite size is comparable to, or larger than the thermal energy. Such grains are too small to become superconducting by themselves. They can be considered as "nano-scale" atoms. Instead of taking apart atoms to induce the M/I transition, one takes apart these quantum dots. On-site repulsion is determined by the charging energy of the grains. Here the control parameter is the inter-gain coupling. In the case of the cuprates, one starts from the insulating state, and one reaches the metallic state by adding carriers ("doping" the insulator). Enhanced screening reduces on-site repulsion, and eventually allows metallic transport. Here the control parameter is the doping level.

A parallel is drawn between the respective phase diagrams of these two kinds of materials, their respective transport properties in the normal state and their superconducting behavior. The role of phase fluctuations of the superconducting order parameter near the M/I transition is emphasized.

4.1 Granular superconductors

Granular superconductors are composed of small grains of a superconducting metal separated by a dielectric barrier. The best known example of a granular superconductor is granular Aluminum, the dielectric barrier being for instance amorphous Al_2O_3 or Ge. Granular Aluminum can be produced by reactive evaporation of Al in the presence of a partial pressure of oxygen, or by co-deposition of the constituents in the case of Al-Ge. The grain size is typically in the nanometer range. The grain size and the inter-grain electrical resistance are the main parameters that control the normal state and superconducting properties. In an idealized model where all grains have the same size d and all inter-grain barriers have the same resistance R, the macroscopic resistivity is given by $\rho = Rd$. In the granular case, it is the conductivity at the grain scale, namely R^{-1}, that controls the distance from the M/I transition.

4.1.1 The granular structure

When atoms from two different elements strike at random the surface of a substrate, and under conditions of deposition such that one of the elements crystallizes and the other remains amorphous, the granular structure is formed. Because metals have a much stronger tendency to crystallize than dielectrics, co-deposition of a metal and of a dielectric will often result in the formation of the granular structure. Such is the case for granular Al, but also for other mixtures such as Ni-SiO_2 (for further reading on granular metals, see the review of B. Abeles referenced below).

The idea proposed to explain the formation of the granular structure is that nucleation and growth of metallic crystallites result in the expulsion of dielectric atoms or molecules towards the periphery of the growing crystallite. When the amorphous dielectric forms a continuous coating around the metallic grain, growth stops. This model explains three key features of granular structures: the uniformity of the grain size, the fact that the thickness of the dielectric is on the atomic scale, and the decrease of the grain size as the inverse of the volume fraction of the dielectric (Fig. 4.2).

This last property results directly from the constant dielectric thickness: the dielectric volume fraction varies as the surface to volume ratio of the grain, i.e. as the inverse of the grain size. This law is observed up to a certain dielectric volume fraction, beyond which the grain size saturates at a certain value, which varies with the temperature of the substrate during deposition. It decreases with the temperature of the substrate. When the substrate is held at room temperature during deposition, the grain size saturates around 30 to 40Å. When it is held at liquid nitrogen temperature, it reaches 20 to 30Å. In the saturated grain size regime, at higher dielectric volume fraction, the dielectric barrier thickens up, which induces the M/I transition. At still higher dielectric volume fraction, the whole structure becomes amorphous.

This growth model of the granular structure is confirmed by experiments showing that when the temperature of the substrate at the time of deposition is high enough for the dielectric to crystal-

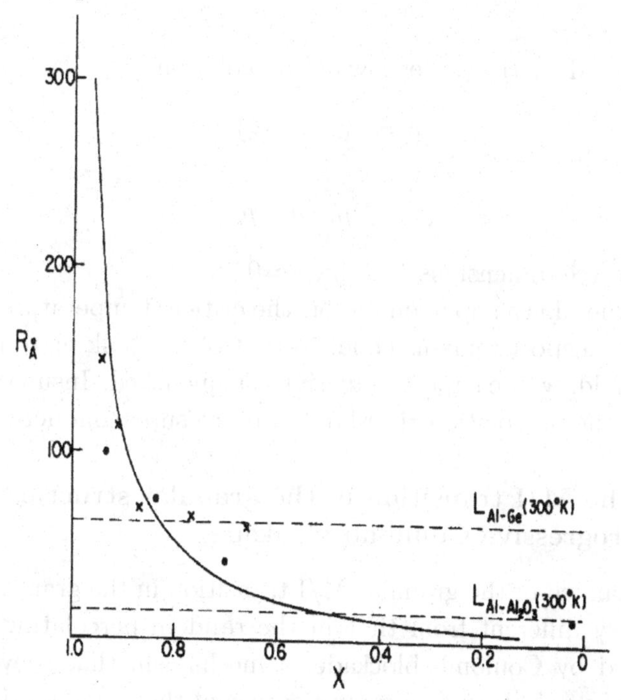

Figure 4.2: Aluminum grain size as a function of dielectric volume fraction in Al-Al_2O_3 films. The continuous curve is a fit to a $1/d$ law. After Y. Shapira and G. Deutscher, Thin Solid Films **87**, 29 (1982).

lize as well as the metal, a completely different structure is formed, characterized by large clusters of crystallites of both constituents. This structure is well described by percolation theory. Its properties are very different from those of the granular structure, in particular the M/I transition is governed by a purely geometrical percolation process. The mixture is metallic as long as there exists an infinite metallic cluster. When that cluster loses its last point of connection, its two parts are separated by a dielectric grain, and not by a thin barrier. Electron tunneling through the dielectric grain is negligible, and the mixture just becomes insulating. Near the threshold, the variation of the conductivity and other transport phenomena such as the Hall effect, as a function of the metal volume fraction, are well described by the power laws of percolation:

$$\sigma = \sigma_0 \left(p - p_c\right)^t \tag{4.1}$$

$$R_H = R_{Ho} \left(p - p_c\right)^{-g} \tag{4.2}$$

where in three dimensions t=2 and g=0.38.

If the metal is a superconductor, the critical temperature as measured by transport remains equal to that of the bulk material up to the threshold, where superconductivity is quenched. Insulating samples, past the percolation threshold, are not superconducting.

4.1.2 The M/I transition in the granular structure: progressive Coulomb blockade

The mechanism of the granular M/I transition in the granular structure is very different from that in the random percolating one. It is governed by Coulomb blockade, a mechanism that prevents the addition of an electron to a grain because of the electrostatic energy involved. Beyond a certain value of the inter-grain resistance, this on-site repulsion becomes effective. The granular material then becomes in effect a mesoscopic Mott insulator, where the on-site repulsion is on the scale of the grain rather than on that of the atom.

To understand how the M/I transition occurs, one needs to compare the charging energy E_c — given by the grain size — to the inverse of the time τ it takes an electron to relax back to its original site (grain). If $E_c < \hbar/\tau$, the electrostatic charging energy is ineffective in blocking electron transfer and the contact is in fact metallic. Because $\tau = RC$, where C is the capacity of the grain, and because both C and E_c^{-1} are proportional to the grain size, the only parameter left is R. The condition for metallicity is simply:

$$R < \frac{\hbar}{e^2} \tag{4.3}$$

and for the conductivity on the macroscopic scale:

$$\sigma > \frac{e^2}{\hbar d} \tag{4.4}$$

which is equivalent to the Mott criterion for metallicity, the grain size replacing here the atomic size. This modified Mott criterion for metallicity is in good agreement with experiment. The sign of the resistivity temperature coefficient changes at about that value (see the review of Abeles in further reading). It is also known that the metallic state does persist up to larger resistivities as the grain size is made larger, in agreement with Eq. (4.4).

The criterion for the threshold that we have used here considers the transfer of one single electron from a grain to its neighbor. In other words, the granular metal behaves at that point as if it had only one free electron per grain — while the bulk metal has typically one per atom. A generalized criterion giving the effective number of free electrons per grain p as a function of R would be:

$$p^2 = \frac{\hbar}{Re^2} \tag{4.5}$$

Near the M/I transition, the Hall constant should then vary as the square root of the inter-grain resistance, or approximately as the square root of the resistivity since in that regime the grain size is nearly constant:

$$R_H \propto \rho^{1/2} \tag{4.6}$$

This behavior has indeed been observed in granular Al (see further reading for references). It is very different from the behavior predicted by percolation theory. From the percolation power laws in 3D:

$$R_H \propto \rho^{0.15} \tag{4.7}$$

The value of the resistivity near threshold and the measured variation of the Hall constant with resistivity show that Coulomb effects do dominate the M/I transition in the granular structure.

4.1.3 Transport regimes and the phase diagram

The role played by Coulomb effects is also apparent when one studies transport as a function of temperature and superconductivity.

The three transport regimes

Three different transport regimes have been identified in the normal state:

(i) for resistivities much smaller than the critical value $\left(\frac{\hbar d}{e^2}\right)$,or about $1{\cdot}10^{-3}\Omega$cm for a typical grain size, the behavior is metallic ($d\rho/dT > 0$), with a weak temperature dependence. Scattering is dominated by grain boundaries. The mean free path is about equal to the grain size. This regime persists up to resistivities of the order of $100\mu\Omega$cm.
(ii) for resistivities near the critical value, the temperature coefficient of resistivity changes sign. One observes a logarithmic divergence at low temperatures:

$$\rho^{-1} \propto \ln T \tag{4.8}$$

This regime extends up to resistivities of the order of $1{\cdot}10^{-2}\Omega$cm (Fig. 4.3).
(iii) for resistivities much larger than the critical value, the divergence is exponential:

$$\rho \propto \exp - \left(\frac{T_0}{T}\right)^{1/2} \tag{4.9}$$

which is typical of an insulator dominated by Coulomb interactions.

By contrast, in the random percolating structure the conductivity is metallic up to threshold and beyond threshold the resistivity increases exponentially. There is no intermediate logarithmic range. The exponential dependence:

$$\rho \propto \exp - \left(\frac{T_0}{T}\right)^{1/4} \tag{4.10}$$

is characteristic of variable range hoping, as seen in amorphous semiconductors. This behavior occurs when transport is governed by the availability of states at adequate energy levels, rather than by Coulomb effects.

The intriguing regime is the logarithmic one, for which there is so far no accepted theoretical interpretation.

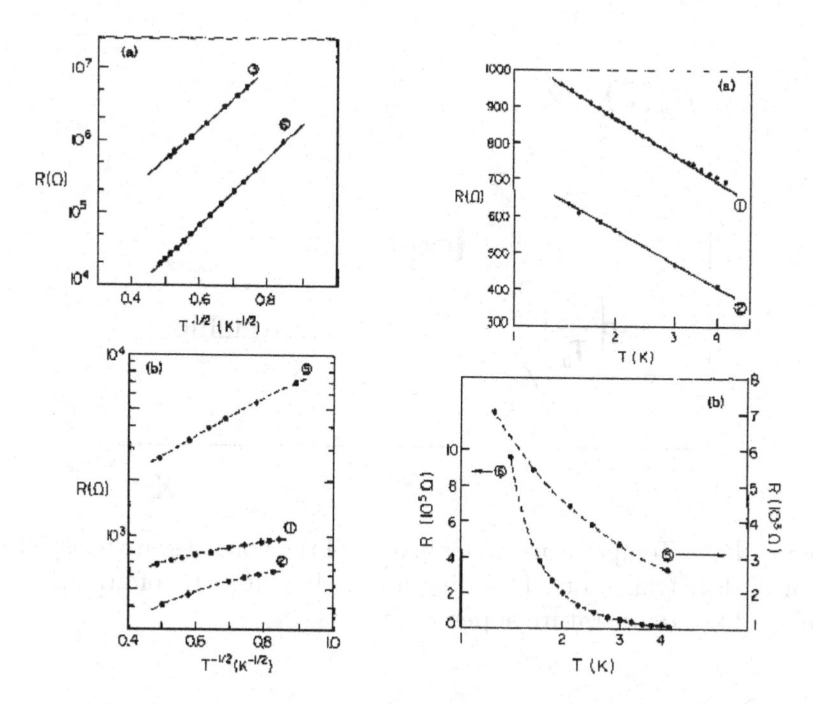

Figure 4.3: Normal state resistivities of granular Al-Al_2O_3. The temperature dependence is exponential for the non-superconducting samples 3 and 6, and logarithmic for superconducting samples 1 and 2. See Table below for sample characteristics. After G. Deutscher, B. Bandyopadhyay, T. Chui, P. Lindenfeld, W.L. McLean and T. Worthington, Phys. Rev. Lett. **44**, 1150 (1980).

The phase diagram

There is a close connection between the different transport regimes and the phase diagram in the (T, x) plane (Fig. 4.4).

The superconducting state exists below a line $T_c(x)$. As the metal volume fraction is reduced, T_c first goes up, reaches a maximum and eventually drops down to zero. As T_c first rises, transport is metallic; around the plateau region and in the decreasing part it becomes logarithmic and when superconductivity is finally quenched, it is exponential. In the insulating regime, the grains may or may not be superconducting, depending on whether their size is above or below

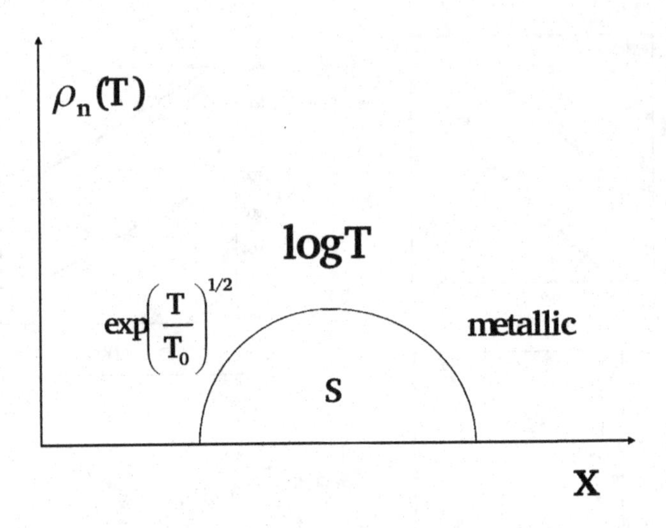

Figure 4.4: Temperature dependences of the normal state resistivity near a Mott transition. This diagram is characteristic of granular as well as high temperature superconductors.

the critical value where the number of Cooper pairs is of order unity (or equivalently where the electronic level splitting is equal to the gap), Eq. (3.15) (see Gerber *et al.* for further reading). If their size is above critical, a "super-insulating" regime is observed, the resistance raising more quickly as the temperature is lowered below the critical temperature of the grains. This behavior has been attributed to the reduced conductance of inter-grain tunneling when the grains are superconducting, because of the reduced density of states near the Fermi level. This "super-insulating" regime is not observed when the grain size is below critical.

The impact of the grain size on the phase diagram is twofold (Fig. 4.5):

(i) The maximum critical temperature goes up as the grain size is reduced. To a good approximation this increase varies as the inverse of the grain size. For a grain size of 30Å, T_c reaches up to more than 3K, an increase of almost a factor of 3 when compared to that of bulk Al.

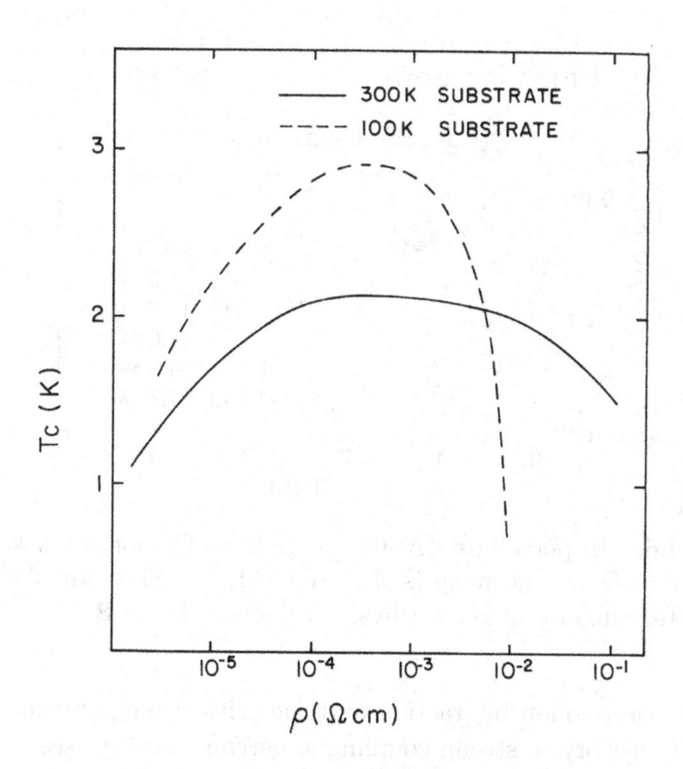

Figure 4.5: Critical temperature of granular Al deposited at room temperature (continuous line) and liqud nitrogen temperature (broken line). Low temperature deposition produces smaller grains sizes, higher maximum T_c and lower resistivity threshold value.

(ii) The maximum value of the resistivity up to which superconductivity persists increases with the grain size.

While the relation between the grain size and the threshold value of the resistivity at the M/I and S/I transition is well explained by the progressive Coulomb blockade model, Eq. (4.4), the origin of the large increase of T_c is still not well understood. No such increase is observed in random percolating structures where there is no Coulomb blockade effect (Fig. 4.6). Various models such as a softening of phonons at the grain's surface, have been reviewed by Abeles (see further reading). A phonon softening does increase

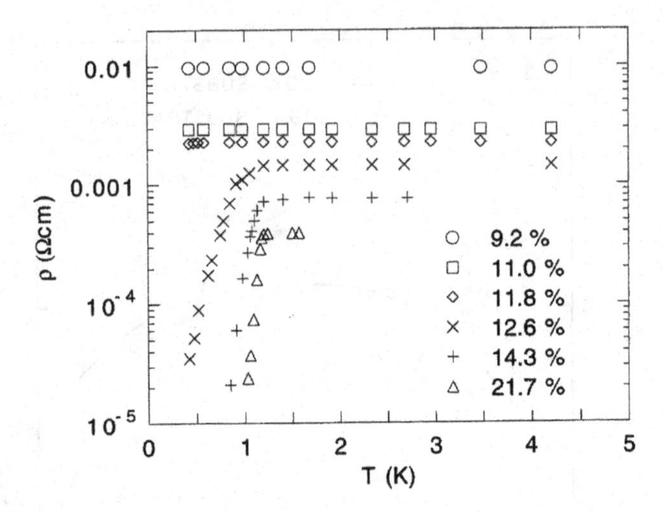

Figure 4.6: In percolating Al-Ge, there is no Coulomb blockade effect and no T_c enhancement is observed. After J. Shoshany, V. Goldner, R. Rosenbaum *et al.*, J. Phys.: Condens. Matter **8**, 1729 (1996).

the electron-phonon interaction and the critical temperature in the McMillan theory of strong coupling superconductivity (see Chapter 8). However there is little direct experimental evidence for such a softening in granular Al.

Mechanism of T_c enhancement

In the conventional theory of superconductivity, the electron-phonon interaction is considered to be favorable for superconductivity because it triggers an indirect electron-electron attraction, while the direct Coulomb electron-electron interaction is unfavorable since it is repulsive. One might therefore have expected that the enhancement of the Coulomb repulsion near the M/I transition of the granular structure, should weaken rather than enhance superconductivity. Since this expectation is contradicted by experiment, one must conclude that the view that the electron-phonon interaction is good for superconductivity, and the direct Coulomb interaction is bad, is far too naive. It is remarkable that T_c enhancement is largest for the smallest grains, for which Coulomb effects are strongest. These observations raise the intriguing possibility that, up to a point, enhanced Coulomb interactions may in fact be favorable for superconductivity. As already noted, when Coulomb interactions are not involved in the M/I transition, such as is the case for the random percolative structure of Al/dielectric mixtures (Al/Ge), there is no increase at all of T_c (Fig. 4.6).

A simple model describing the direct and indirect electron-electron interactions considers a mixture of positive ions and electrons, in which the ions are treated as a fluid rather than a solid. It is called the Jellium model, first studied by Nozieres and Pines. The electron-electron interaction comprises two terms. The first one is the direct repulsive screened Coulomb interaction, the second one the indirect interaction mediated by the ions vibrations:

$$V(q,\omega) = \frac{4\pi e^2}{q^2 + k_s^2} + \frac{4\pi e^2}{q^2 + k_s^2} \frac{\omega_q^2}{\omega^2 - \omega_q^2} \tag{4.11}$$

where k_s^{-1} is the Thomas Fermi screening length given by:

$$k_s^2 = \frac{6\pi n e^2}{E_F} \tag{4.12}$$

n being the electronic density. The frequency ω_q describes the dispersion of ions vibrations, which is limited by the Debye frequency ω_D corresponding to the Debye wave length q_D. This models does not take into account the retarded nature of the attractive part of the electron-phonon interaction, which is why at zero frequency the interaction potential is zero. But the important point to be learned from the Jellium model is that the two terms of the interaction are in fact closely related. They can both be increased by decreasing the carrier density. In 3D, this does not lead to an enhanced critical temperature, because the decrease of the density of states at the Fermi level will more than compensate this increase. But in 2D, it will, because the density of states is independent from the carrier density (for further reading see G. Deutscher).

The Jellium model (besides giving the right order of magnitude of the interaction potential) draws our attention to the fact that the

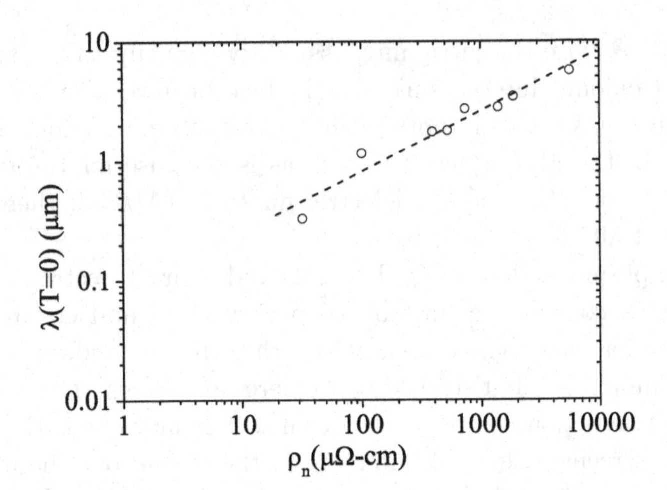

Figure 4.7: Increase of the penetration depth in granular Al-Al_2O_3 near the M/I transition. After D. Avraham *et al.*, J. de Physique Colloque C **6**, 586 (1978).

(generally accepted) dissociation between repulsive and attractive interactions is actually artificial and misleading. The net attractive interaction may be enhanced by increasing the repulsive part.

One should also keep in mind the possibility that highly degenerate harmonics in nearly spherical grains can give rise to an enhanced density of states at the Fermi level. This is a geometrical effect similar in a sense to the enhanced density of states near a van Hove singularity, discussed in some detail in Chapter 8.

4.1.4 Loss of superfluid density and decrease of T_c near the M/I transition

The superconducting regime near the M/I transition, where the grains become progressively decoupled, is of particular interest. This regime is characterized by a severe loss of superfluid density. The penetration depth near the M/I transition in Al/Al_2O_3, increases as the square root of the resistivity (Fig. 4.7).

The increase is spectacular. The penetration depth reaches values of up to several micrometers, while its value in pure Al is 350Å. The superfluid density is reduced by a factor of up to $1{\cdot}10^4$. This large reduction is consistent with our picture of a progressive Coulomb blockade. Near the transition, the *effective* number of free electrons per grain is of order unity, while in a 30Å grain the number of conduction electrons is of the order of a few 10^2. This fits well with the measured loss of superfluid density: near the M/I transition, there is only on the order of one free electron per grain. In addition, itinerant electrons have a strongly enhanced effective mass. Both the reduced effective carrier density and their increased effective mass contribute towards the strong enhancement of the penetration depth.

At such small superfluid densities, phase fluctuations of the order parameter become important. As already discussed in Chapter 2, Eq. (2.38), phase gradients destroy the coherence of the superconducting state and limit the value of the critical (coherence) temperature to:

$$k_B T_c = A \frac{\Phi_0^2}{\lambda^2} d \tag{4.13}$$

where $A{=}(15\pi^3)^{-1}$, and we have replaced the coherence length by the grain size. When the penetration depth is small and the coherence length large, as is the case in conventional superconductors, this coherence limit for T_c falls well above the critical temperature of the bulk material. But when it is very large, and the coherence length (or the grain size in our case) is short, it can fall below the bulk T_c. It then becomes effectively the critical temperature of the sample.

This is exactly what happens near the M/I transition in granular Al. Plugging into Eq. (4.13) the largest penetration depth (5 microns) measured for a sample having a resistivity of $1{\cdot}10^{-2}\Omega$cm, and the grain size of 30 Å, we obtain a transition temperature of about 1K, a remarkable agreement between theory and experiment, obtained without any adjustable parameter, which demonstrates that phase fluctuations can indeed determine the critical temperature near an M/I transition. At a resistivity of $1{\cdot}10^{-3}\Omega$cm, the superfluid density is already large enough not to affect the transition temperature. The transition is still basically a BCS one, in agreement with heat capacity data, see Fig. 4.8 below.

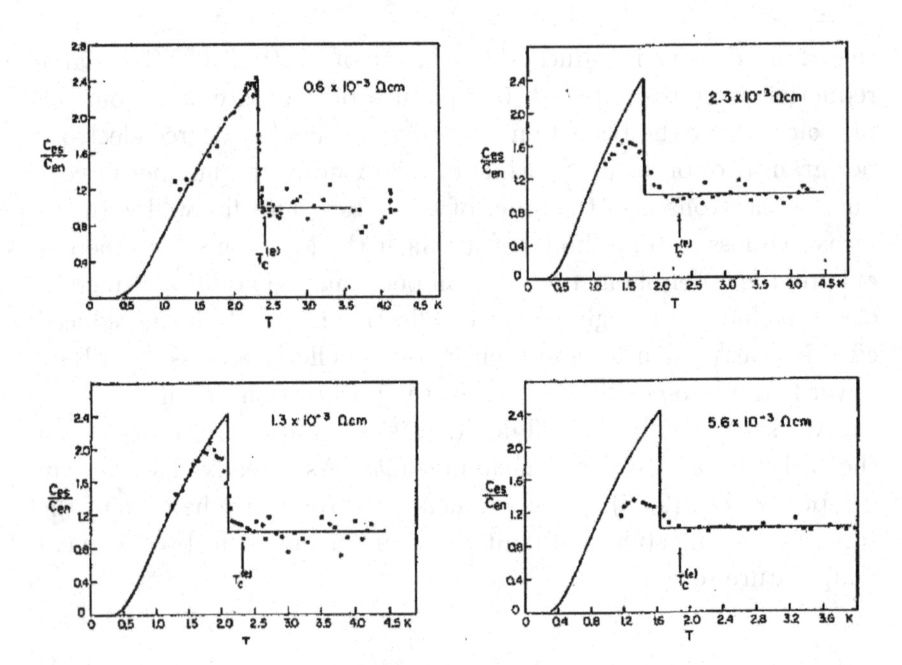

Figure 4.8: Heat capacity transitions for granular Al-Al_2O_3 films. For resistivities lower than $1{\cdot}10^{-3}\Omega$cm, there is a good fit to BCS, but at higher values the transitions become similar to those in underdoped YBCO, see proceeding chapter. After T. Worthington *et al.*, Phys. Rev. Lett. **41**, 316 (1978).

There is here an interesting connection between a small superfluid density regime and Bose–Einstein condensation. In both cases superfluid condensation is achieved through phase locking, while in the BCS condensation the formation of Cooper pairs (bosons) occurs at the same time as phase locking.

4.1.5 The short coherence length of granular superconductors

One can guess that if there is about one Cooper pair per grain near the M/I threshold, and the grains are weakly coupled as is the case near the threshold, the coherence length must then be of the order of

Table 4.1: Table for samples shown Fig. 4.3.

Specimen number	ρ_{rt} (Ω cm)	$\rho_{4.2}$ (Ω cm)	ℓ (μm)	T_c K	Form or R_N vs T
1	0.04	0.16	6	1.9	Logarithmic
2	0.02	0.12	10	1.8	Logarithmic
5	0.02	0.10	1	1.2	
6	0.05	3	1	...	Exponential
3	0.3	32	1	...	Exponential

the grain size. This is very short compared to the coherence length typical of low temperature superconductors.

In the dirty limit of the metallic regime, where the mean free path l is much shorter than the clean limit value ξ_0=00.18 $(\hbar v_F/k_BT_c)$, the coherence length is given by ξ =0.85$(\xi_0 l)^{1/2}$. The value of the mean free path in a metal can be obtained from the known value of the product $\rho l \approx 1{\cdot}10^{-11}\Omega\text{cm}^2$. If we take ρ=$1{\cdot}10^{-4}\Omega$cm, near the limit of the metallic regime, we obtain l =$1{\cdot}10^{-7}$cm. In Al, $\xi_0 \approx 1{\cdot}10^{-4}$cm, and we obtain $\xi \approx$250Å, larger than the grain size (which is less than 100Å). On the scale of the coherence length, the superconductor is homogeneous.

At the upper end of the intermediate transport regime where the resistivity has a logarithmic divergence at low temperatures but the sample is still superconducting, we get with ρ=$1{\cdot}10^{-2}\Omega$ cm (see Table 4.1) a mean free path of 0.1Å. This value of course does not make much sense: the granular material is not a metal anymore. It only expresses the fact that grains are very weakly coupled. Yet, if we use this mean free path value to calculate the coherence length, we get this time a value of about 30Å, just about the grain size. Grains are effectively on the verge of being decoupled from the standpoint of superconductivity. In view of their small size, we expect large fluctuations effects, as indeed observed for instance in the heat capacity transition. The crossover from homogeneous to granular behavior is evident in the heat capacity data (Fig. 4.8).

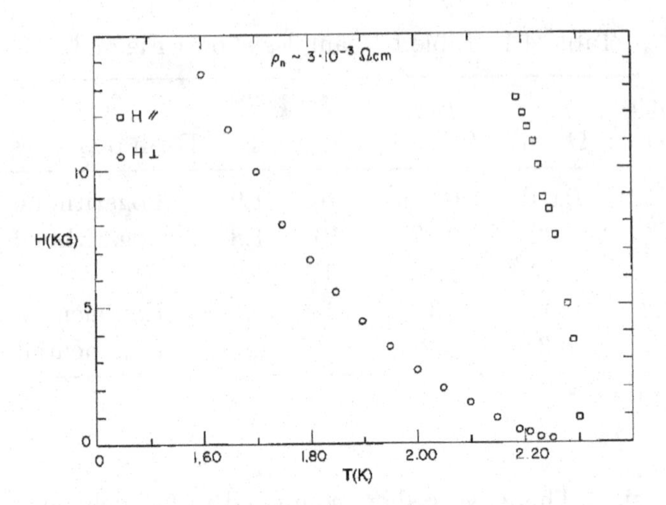

Figure 4.9: Critical fields of a relatively high resistivity granular Al film. Note the strong upward curvature in the perpendicular field orientation, similar to the irreversibility field in High T_c cuprates. After G. Deutscher and S. Dodds, Phys. Rev. B **16**, 3936 (1977).

As we have seen in Chapter 2 the value of the coherence length can be determined from a measurement of the upper critical field $H_{c2} = (\Phi_0/2\pi\xi^2(T))$. In the metallic regime, far from the M/I threshold, $H_{c2}(T)$ is a linear function of the temperature near T_c, as predicted by the Ginzburg–Landau theory. The value of the coherence $\xi(0)$ obtained from the extrapolation of H_{c2} to low temperatures is larger than the grain size, and in agreement with that calculated from the normal state resistivity obtained (see Abeles for further reading). Although granular, the film behaves as if it were homogeneous.

In the intermediate regime, closer to the threshold, a different behavior is observed. In the perpendicular orientation, resistive behavior occurs above a field $H^*_{c2}(T)$ that shows a marked upward curvature (Fig. 4.9).

A strong anisotropy is observed, the transition taking place at a much higher field in the parallel orientation. The origin of this anisotropy and of the nonlinear behavior in the perpendicular orientation is not fully understood. It may result from a complicated interplay between the destruction of inter-grain coherence near T_c where $\xi(T)$ is larger than the grain size and of intra-grain superconductivity at lower temperature where the coherence length becomes of the order of the grain size. Extrapolation to low temperatures gives $H_{c2}(0) \approx$8T, $\xi(0) \approx$60Å, not far from the grain size.

Alternatively, one may remark that the resistive transition in a field determines in principle the irreversibility field, i.e. the field beyond which vortices are not pinned anymore, rather than the upper critical field. These two fields are not in general identical, see Chapter 7 for a detailed discussion of this point. In the perpendicular orientation, a Lorentz force is applied to vortices which tends to move them in the direction perpendicular to the current, which induces a longitudinal voltage. A low resistive transition field in that orientation may just mean that vortex pinning is very weak. Indeed, this is typically a short coherence length effect, also observed in the cuprates. There, weak pinning is ascribed to the small value of the condensation energy per coherence volume, compared to the thermal energy. In the granular case, the origin of the small value of this energy is obvious in the intermediate regime where as we have seen the coherence length is of the order of the grain size. The condensation energy per coherence volume is then essentially the condensation energy per grain, itself of the order of $k_B T_c$ for Al grains of about 50A. The observed upward curvature of the irreversibility field is due to weak pinning resulting from strong thermodynamical fluctuations, see Chapter 7 for a detailed discussion of this point.

Very near threshold, destruction of intra-grain superconductivity dominates at all temperatures. The low temperature critical field is then set by the Pauli limit where $\mu_0 H$ =0.7Δ. This limit is also important for the cuprates, see Chapter 7 for more details.

4.2 Phase diagram of the cuprates

Cuprates consist of weakly coupled CuO_2 planes. Superconductivity appears only upon hole (or electron) doping of these planes. The

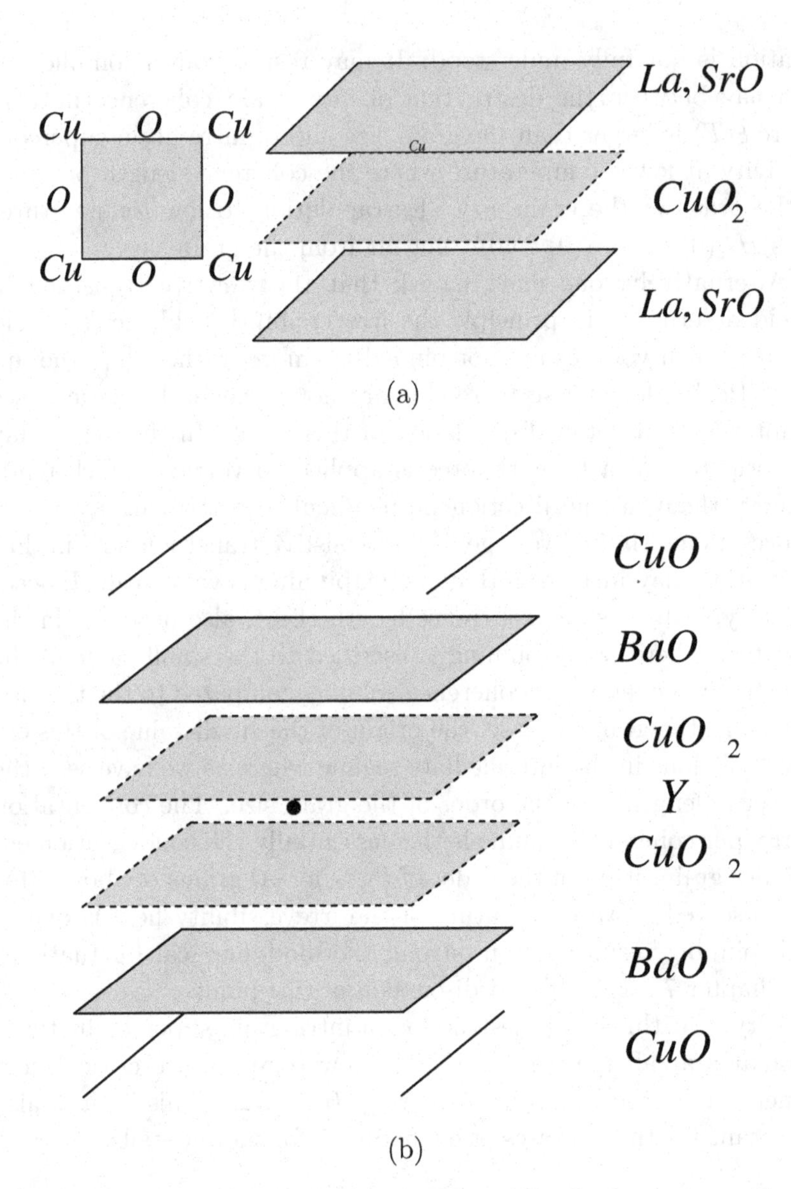

Figure 4.10: Schematic structures of $La_{2-x}Sr_xO_4$ (upper panel) and $YBa_2Cu_3O_7$ (lower panel). LaSrO planes and CuO chains serve respectively as charge reservoirs for CuO_2 planes.

undoped compound is an antiferromagnetic insulator. Two examples are $La_{2-x}Sr_xCuO_4$ with a unit cell composed of one layer of CuO_2 intercalated between two layers of La,Sr,O (Fig. 4.10(a)), and $YBa_2Cu_3O_7$ (Fig. 4.10(b)), composed of two layers of CuO_2 separated by a layer of Y and sandwiched between layers of BaO and chains of CuO.

Valence counting shows that La_2CuO_4 (214) and $YBa_2Cu_3O_6$ (123) have one electron per unit cell. According to band theory they should be metallic. Their antiferromagnetism is generally attributed to strong electron-electron Coulomb repulsion, which does not allow two electrons to be on the same Cu site at the same time. This is the Mott insulator model (another model that has been proposed for the cuprates is that of a band anti-ferromagnetic insulator, see Chapter 8 for details on this model). In an ideally uniform granular structure with an uneven number of electrons per grain, one would also expect to have an antiferromagnetic order in the weak coupling limit.

While metallicity is induced in the granular structure by increasing inter-grain coupling, a metallic state is induced in the cuprates by hole or electron doping. In the 214 compound this is achieved by replacing part of the La (3+ ion) by Sr (2+ ion); and in 123 by adding oxygen in between the Cu atoms that form the chains. When chains are filled, the 0_7 concentration is reached. Chains play a double role: they act as charge reservoirs for the CuO_2 planes, and because they are metallic in their own right they provide a good coupling between neighboring stacks of CuO_2 planes, thereby reducing the anisotropy.

Doping destroys the antiferromagnetic state and beyond a certain level induces superconductivity. As a function of the chemical doping p counting the number of doped holes per CuO_2 layer per unit cell, superconductivity appears at $p = p_C$, passes through a maximum at $p = p_M$, and then decreases, vanishing at a concentration p_F. In the compound 214, p_M=0.16 and p_F=0.3. In 123, the doping is counted in terms of the total oxygen concentration (between 6 and 7). There is experimental evidence that the Cu chains become themselves metallic when they are sufficiently filled with oxygen, therefore the counting in terms of doped holes per CuO_2 layer per unit cell is

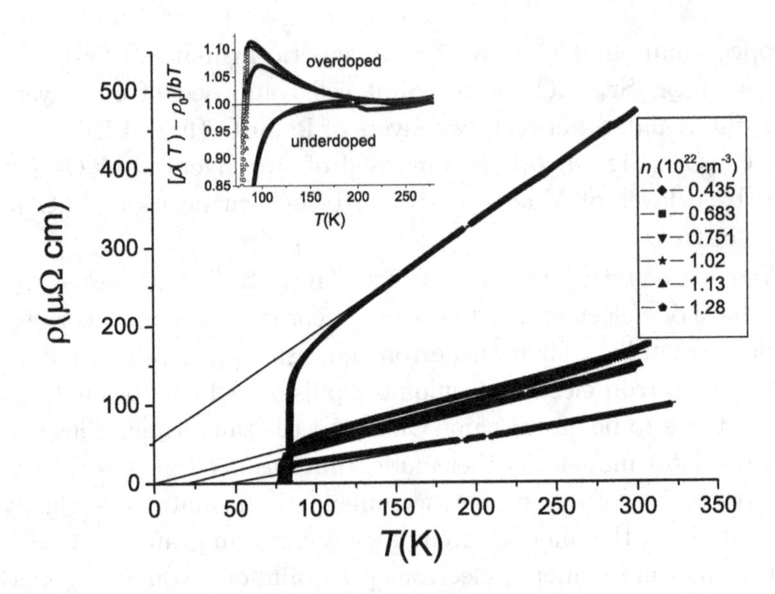

Figure 4.11: Resistivities of YBCO films bend downward at a temperature T^* in underdoped samples and upward at a temperature T_f in overdoped ones. After H. Castro and G. Deutscher, Phys. Rev. B **70**, 174511 (2004).

not obvious. A maximum T_c of 92/93K is reached for $0_{6.93}$, it reduces to 87 to 88K at 0_7. The effect of atomic disorder on superconductivity in the cuprates is by itself an issue, particularly because of the d-wave symmetry of the order parameter, see next chapter. In general, doping and disorder are not independent from each other. In 214, doping increases disorder since Sr atoms are substituted in the La planes, while in 123 disorder is reduced. The composition 0_7 is thought to give the most perfect crystalline order, because there are no vacancies left on the Cu chains.

4.2.1 Transport regimes and the phase diagram

The nature of the normal state above the $T_c(p)$ line, outside of the superconducting dome, depends on the doping level. In hole doped cuprates, this is reflected inter alia in distinct transport regimes (Fig. 4.11).

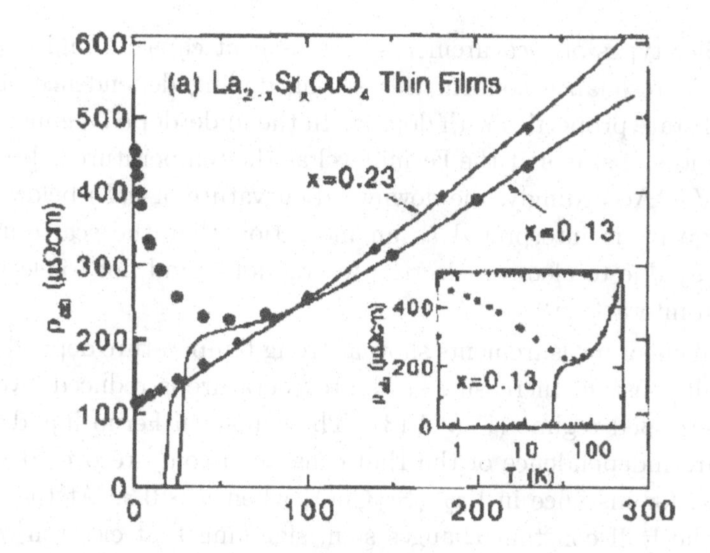

Figure 4.12: The normal state resistivity of slightly underdoped $La_{2-x}Sr_xCuO_4$ films (x =0.13) rises logarithmically at low temperatures. Note the similarity with the behavior of granular Al films, see Fig. 4.3. After Ando *et al.*, Phys. Rev. B **56**, R8530 (1997).

(i) In overdoped samples $p > p_M$, the temperature dependence of the resistivity is typical of a metal, with $\rho(T)$ showing a positive curvature below a temperature T_f and a linear dependence above it.
(ii) At exactly $p = p_M$, $\rho(T)$ is linear from above room temperature down to T_c, and extrapolates to zero at T=0. This remarkable behavior is thought to be a key feature of the superconducting cuprates.
(iii) In underdoped samples $p_c < p < p_M$, $\rho(T)$ is linear at high temperatures and shows a downward curvature below a temperature T^*. It increases as $\ln(T)$ at low temperatures $T<T_c$ (Fig. 4.12).
(iv) $\rho(T)$ increases exponentially as $\exp\left(-\frac{T_0}{T}\right)^{1/2}$ below p_c.

One will have noticed the similarity with the different transport regimes in granular superconductors, in particular the logarithmic and exponential regimes.

Other types of measurements such as heat capacity and nuclear magnetic resonance confirm this strong doping dependence of the normal state properties with doping. In the underdoped regime, they show a loss of states at the Fermi level as the temperature is lowered below T^*. Accordingly, the downward curvature of $\rho(T)$ below that temperature is interpreted as an indication that the scattering is primarily due to electron-electron interactions (and not to electron-phonon interaction).

Hall effect measurements show a strong temperature dependence, the Hall constant increasing as the temperature is reduced, even in the overdoped regime (Fig. 4.13). The canonical Fermi liquid temperature independence of the Hall constant is only recovered when $p > p_F$, for instance in $La_{2-x}Sr_xCuO_4$ when $p > 0.3$. At the same time, the Hall constant changes sign, signaling that electron orbits go from hole-like to electron-like. In superconducting samples, the temperature dependence of the Hall constant above T_c is the least pronounced at optimum doping $p = p_M$. Only after superconductivity is quenched does one apparently recover the fully normal metallic behavior.

In a conventional metallic superconductor, one does not need to have a detailed knowledge of the normal state properties to predict the main superconducting properties. The only normal state parameter one needs to know is the mean free path. On the contrary, in the case of the cuprates, normal state properties are considered as possibly giving important clues regarding the origin of the High T_c mechanism itself. Of particular importance is the loss of states at the Fermi level below the temperature T^* in the underdoped regime, below which the spin susceptibility is reduced and the resistivity presents a downturn. It has been interpreted as a temperature where a spin gap, or a pseudo-gap opens up. The opening of this pseudo-gap might be a precursor of superconductivity, signaling the appearance of a pairing amplitude, superfluid coherence being only achieved at T_c. This would be similar to the granular case where as we have discussed above pairing appears in the grains before inter-grain phase are locked. A more detailed discussion of the pseudo-gap is given in the next chapter.

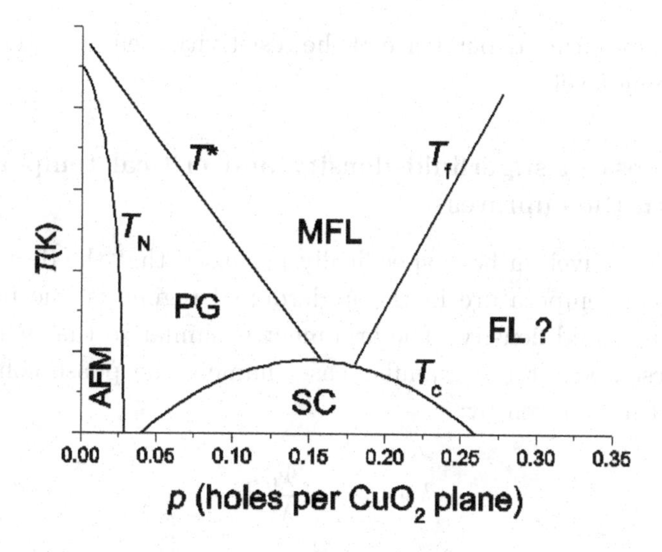

Figure 4.13: Main domains in the phase diagram of hole doped cuprates.

The scaling temperature T_H below which the Hall effect temperature dependence becomes important (Fig. 4.13) is very close to the temperatures T^* and T_f below which the resistivity ceases to be linear in temperature, respectively in the underdoped and overdoped regimes. It has a symmetrical behavior around optimum doping, where it is at a minimum. Its increase on the underdoped side is not surprising, as it may just reflect the loss of states below a pseudo-gap opening temperature. It is more surprising on the overdoped side, usually considered as approaching progressively a normal metallic state, in which case the Hall constant should become less and less temperature dependent as overdoping is increased. Instead, the doping behavior of T_H^* is non-monotonous. This is the reason why samples in the overdoped regime have been called Anomalous Fermi liquids.

The fact that the temperature dependence of the Hall constant is at its minimum when T_c is at its maximum may be of importance. It confirms that optimum doping is not just a crossover, but a special concentration in the phase diagram, as already apparent from the linear temperature dependence of the resistivity seen at, and only at that doping level.

4.2.2 Loss of superfluid density and critical temperature in the cuprates

Emery and Kivelson have specifically proposed that the decrease of the critical temperature in the underdoped regime is due to a reduced superfluid density. The argument is similar to that which we have presented for the granular case, namely the phase coherence temperature is given by:

$$k_B T_{co} = A \frac{\Phi_0^2}{\lambda^2} d \tag{4.14}$$

where d, the inter-layer spacing, replaces the grain size (Eq. 4.11) or the coherence length in the original formulation of Chakraverty and Ramakrisnan. The interpretation of the phase diagram proposed by Emery and Kivelson is that below optimum doping the coherence temperature T_{co} becomes lower than the temperature at which a pairing amplitude appears (here the pseudo-gap temperature) and becomes effectively the critical temperature of the cuprate, while in the overdoped regime the decrease of T_c is really due to a decrease of the interactions causing superconductivity.

Indeed, in the underdoped regime the critical temperature varies linearly with the λ^{-2}, as first observed by Uemura who interpreted this behavior as indicating a BE condensation. The two points of view are of course closely related, as in both cases superfluidity is achieved by phase locking, just as it is in the granular case.

As also reported by Uemura, the superfluid density decreases in general again in the overdoped regime, although the carrier density increases. This is the so-called "boomerang" effect. In fact, the linear relationship between the critical temperature and the superfluid density is in general similar on both sides of the phase diagram. It is therefore not clear whether the decrease of the critical temperature on the overdoped side is due to a weakening of the interactions leading to superconductivity, or as on the underdoped side just to a loss of superfluid density. The origin of this loss on the overdoped side is itself not clear. It may be due to an increased disorder, or to a more fundamental increase of the scattering rate. Indeed, the temperature T^*_{RH} below which the Hall constant is temperature dependent (increasing as the temperature is reduced), goes up in the overdoped side, as well as on the underdoped one, as discussed above. Increased scattering on both sides of the diagram, whatever its origin, may be at the origin of the (almost) symmetrical reduction of the superfluid density on both sides of the diagram.

4.3 Summary

Superconductivity near a Mott transition is characterized by a superconducting dome that extends from the vicinity of the insulating state up to the metallic one. Granular and High T_c superconductors are two examples of systems that present such a dome. Superconductivity appears at some distance x_c from the insulating phase, reaches a maximum at $x = x_M$ and decreases for $x > x_M$. Here x is some parameter that characterizes the distance from the insulator.

Transport properties in the normal state vary strongly across the phase diagram:

(i) Outside of the superconducting dome on the insulating side, strong (exponential) localization is observed.
(ii) A new transport regime, characterized by a logarithmic increase of the resistivity at low temperatures, occurs in the range $x_c < x < x_M$, both in granular and High T_c superconductors. This regime is so far not well understood.
(iii) At $x > x_M$, transport is metallic.

In granular superconductors the Hall constant varies as the square root of the resistivity, which is consistent with a progressive Coulomb blockade as one goes from the metallic to the logarithmic regime.

The Hall constant is temperature dependent. In the cuprates, the temperature dependence is at a minimum at $x = x_M$. Together with the linear temperature dependence of the resistivity seen at that point, this minimum shows that it is of special significance.

4.4 Further reading

For a review on granular metals, see B. Abeles, Adv. Phys **24**, 407 (1975); also B. Abeles, Applied Solid State Science **6**, 1 (1976).

For a review of superconductivity in nano-structures, see G. Deutscher in "Physics of Superconductors", Vol. II, p. 25 Eds. K. Benemann and J. B. Ketterson, Springer 2004.

For the transition to the "Super-Insulator" state, see Gerber *et al.*, Phys. Rev. Lett. **78**, 4277 (1997).

For a discussion of the Jellium model, see P.G. de Gennes, "Superconductivity of Metals and Alloys", Benjamin, New York 1966, and original references therein.

For a recent article on normal state transport properties in the cuprates, see H. Castro and G. Deutscher, Phys. Rev. B **70**, 174511 (2004).

Chapter 5

Gap, symmetry and pseudo-gap

The density of states of a superconductor is characterized by an energy scale Δ, over which it is lower than that in the normal state. In the case of conventional superconductors, this scale is basically isotropic around the Fermi surface and is known as the s-wave energy gap. There are no states available in the range $(E_F - \Delta) < E < (E_F + \Delta)$. In the case of the cuprates, Δ is strongly anisotropic, going to zero at nodal points in k space where it reverses sign. This is the d-wave gap. The concept of an energy scale characteristic of the superconducting state is well defined as long as that scale is quite distinct from and much smaller than the kinetic energy of electrons in the normal state, the Fermi energy. This is the case of low temperature superconducting metals or BCS superconductors, where the Fermi energy is larger than the gap by several orders of magnitude. One is then in the weak coupling limit. But it is not in general the case for high temperature superconductors, where the Fermi energy is barely one order of magnitude larger than the gap.

We start by recalling some of the basic properties of the s-wave gap, before turning to a brief review of the experimental evidence for a d-wave gap in the cuprates. The more advanced topic of the pseudo-gap observed in the underdoped regime is then discussed in some detail, including the issue of coherence in the pseudo-gap regime, one of the hot topics of controversy at this time.

5.1 The BCS s-wave gap

In a normal metal, the energies of electron and hole excitations around the Fermi level are given by:

$$\begin{aligned} \epsilon u &= \left(\frac{\hbar^2 k^2}{2m} - \mu \right) u \\ \epsilon v &= -\left(\frac{\hbar^2 k^2}{2m} - \mu \right) v \end{aligned} \tag{5.1}$$

where ϵ is the energy of the excitation measured from the Fermi level, u and v respectively the amplitudes of the electron and hole wave functions, μ is the chemical potential and k the amplitude of the wave vector of the excited state. For an electron excitation $k > k_F$ and for a hole $k < k_F$, where k_F is the Fermi wave vector related to μ by $\frac{\hbar^2 k_F^2}{2m} = \mu$. The signs ensure that the excitation energy is positive for both electron and hole excitations.

5.1.1 The BCS density of states

In the superconducting state, excitations whose energy is close to the Fermi level result from the destruction of Cooper pairs. As shown by Bogoliubov, the above equations for electron and hole excitations are then coupled over the energy range Δ characteristic of the superconducting state:

$$\begin{aligned} \epsilon u &= \left(\frac{\hbar^2 k^2}{2m} - \mu \right) u + \Delta v \\ \epsilon v &= -\left(\frac{\hbar^2 k^2}{2m} - \mu \right) v + \Delta u \end{aligned} \tag{5.2}$$

Solving this set of equations for ϵ we get:

$$\epsilon^2 = \Delta^2 + \left(\frac{\hbar^2 k^2}{2m} - \mu \right)^2 \tag{5.3}$$

where ϵ is now the energy of an excitation consisting of an electron ($k > k_F$) and a hole ($k < k_F$) resulting from the destruction of a

Cooper pair. With:

$$\xi_k^2 = \left(\frac{\hbar^2 k^2}{2m} - \mu\right)^2$$

Eq. (5.3) can be rewritten as:

$$\epsilon_k^2 = \Delta^2 + \xi_k^2 \tag{5.4}$$

This is a basic result of the BCS theory. At the Fermi wave vector, excitations have the finite energy Δ, the energy gap. The excitation energy returns to that characteristic of the normal state, ξ_k, at wave vectors larger than $k_F + \delta k$, where $\delta k^{-1} = (\hbar v_F/\Delta)$. One will have recognized that δk^{-1} is the superconducting coherence length. Knowing the wave vector dependence of the excitation energy, one can calculate the density of states in the superconducting state, $N_S(\epsilon) = N(0) \mid d\xi/d\epsilon \mid$:

$$N_S(\epsilon) = N(0)\frac{\epsilon}{\sqrt{\epsilon^2 - \Delta^2}} \tag{5.5}$$

where $N(0)$ is the normal state density of states at the Fermi level. Here it has been assumed that the normal state density of states is constant within an energy range larger than Δ around the Fermi level, an assumption justified if $\Delta \ll E_F$.

The ratio of the amplitudes of the wave functions of the electron-like and hole-like parts of the excitation is given by:

$$\frac{u}{v} = \frac{\Delta}{\sqrt{\Delta^2 + \xi_k^2} - \xi_k} \tag{5.6}$$

For $\xi_k \gg \Delta$, or $k \gg k_F + \delta k$, excitations have mostly an electron character; for $\xi_k \ll -\Delta$, or $k \ll k_F - \delta k$, they have mostly a hole character. At $k = k_F$, excitations are a mixture of electrons and holes in equal proportion. One can calculate the value of the amplitudes with the additional condition:

$$u^2 + v^2 = 1 \tag{5.7}$$

5.1.2 Giaever tunneling and Andreev–Saint-James reflections

The value of the BCS gap and more generally the density of states in the superconducting state can be obtained from a measurement of the dynamical conductance of weak contacts between the superconductor and a normal metal. Weak contacts can be realized in two different ways: either by introducing a thin dielectric between the superconductor and the normal metal, in which case one has a Giaever tunnel junction; or by making the contact through a constriction that is so small that conduction is limited by the small number of quantum channels, in which case one has a Sharvin contact. Additionally, if the size of the contact is smaller than the coherence length, the superconducting state is unaffected by the contact.

Giaever tunneling

The oxidation of some metals proceeds in such a way that it is self-limiting. This is the case of Aluminum, which forms at or near room temperature a nanometer thick layer of amorphous Al_2O_3. Further oxidation through oxygen diffusion is prevented by a built-in electric field. As shown by Giaever, in spite of the fact that the oxide layer is very thin, it can be pinhole free over the convenient size of a square millimeter. Its conductance is governed by electron tunneling through the barrier, whose height is of the order of the electron volt.

The theory of electron tunneling in superconductors has been developed by McMillan and Rowell (see further reading). When coupling between the electrodes is sufficiently weak, tunneling can be described by an effective Hamiltonian that describes the transfer of an electron in a state of momentum k from the left-hand side into a state of momentum p in the right-hand side:

$$H_T = \sum_{kp} T_{kp} c_k^+ c_p + c.c \tag{5.8}$$

the transition probability per unit time from the state k to the state p is given by

$$W = \left(\frac{2\pi}{\hbar}\right) N(p) \mid T_{kp} \mid^2 \tag{5.9}$$

where $N(p)$ is the density of states in the right-hand side, given by:

$$N(p) = m\frac{L_R}{\pi v_{xR}} \tag{5.10}$$

where L_R is the thickness of the right-hand side electrode and v_{xR} its electron velocity in the direction perpendicular to the junction on that side.

On the other hand, the transition probability is equal to the number of attempts per unit time $v_{Lx}/2L_L$ times the probability of success τ, from which it follows that:

$$\mid T_{kp} \mid^2 = \frac{\hbar}{2m}\frac{v_{Lx}v_{Rx}}{L_L L_R}\tau \tag{5.11}$$

When both electrodes are in their normal state, the density of states of the right-hand side electrode cancels out from the transition probability. It is not possible to "calibrate" it from the density of states of the left-hand side. But in the superconducting state it does not cancel out because the DOS is the product of the normal state density of states (which cancels out), times $\frac{\epsilon}{\sqrt{\epsilon^2-\Delta^2}}$ when $\epsilon > \Delta$ (Eq. (5.6)), and is equal to zero when $\epsilon < \Delta$

At zero temperature, where the Fermi function is infinitely sharp, the derivative of the tunneling current with respect to the applied potential is given by:

$$\left(\frac{dI}{dV}\right)_S = \left(\frac{dI}{dV}\right)_N \left(\frac{\epsilon}{\sqrt{\epsilon^2 - \Delta^2}}\right) \tag{5.12}$$

More generally, when strong coupling effects are taken into account, the gap is a complex function of energy and the normalized dynamical conductance is equal to Re $\left(\frac{\epsilon}{\sqrt{\epsilon^2-\Delta^2(\epsilon)}}\right)$.

The bias dependence of the conductance of a Giaever junction can be qualitatively understood in the following way. At bias values $| eV | < \Delta$, no states are available in S for the incoming electron coming from N: it can only be reflected into N, the conductance is zero. For $| eV | \geq \Delta$, the incoming electron can tunnel into a state that has partial electron and hole character. For $| eV | \gg \Delta$, it tunnels into a state that has an almost pure electron character, and the conductance returns to its normal state value. The conductance has sharp maxima, called coherence peaks, at $| eV | = \Delta$, when tunneling occurs into a state that has equal electron and hole character. When observed, these peaks give conclusive evidence that S is a BCS superconductor. Tunneling from a metal into an insulator will also show a bias range of zero conductance, but without these coherence peaks. The main effect of a finite temperature is to smear these peaks, but they remain very pronounced as long as $T \ll T_c$.

Andreev–Saint-James reflections

Suppose now that there is no dielectric layer between N and S, but instead the conductance of the contact is limited by the size of the contact. It is then equal to the number of quantum channels in the contact area, times the universal conductance $e^2/\hbar$. The number of channels is of the order of $(k_F d)^2$ where d is the size of the contact. When the contact size is smaller than the mean free path, electronic motion through the constriction is ballistic. It is then called a Sharvin contact. In order to describe the electronic wave functions in the vicinity of the contact, one writes that the electron and hole wave functions, and their derivatives, are continuous at the interface. For that purpose one uses the equations describing the excitations on both sides, as we have written them above (Eq. (5.2)).

The essential feature of the solution is that electron and hole wave functions in N become coupled through the pair potential Δ in S. The solution was derived independently by de Gennes and Saint James, and by Andreev (see further reading for references). Just as in the case of a Giaever junction, an incoming electron from N of wave vector k having an energy within the range Δ from the

Fermi energy cannot propagate in S. It can however be reflected as a hole of equal wave vector (and thus of opposite velocity), while a Cooper pair propagates in S. The charge crossing the interface is $2e$ per event. We call these reflections Andreev–Saint-James (ASJ). For the electron-hole coupling in N through Δ to be effective, it is clear that the N/S interface must be highly transparent. When the interface is completely transparent, all electrons incoming from N are reflected in this way, or in other words none undergoes the usual specular electron-electron interface reflection. Since for each scattering event a charge of $2e$ crosses the interface, the conductance is twice as large as in the normal state, or twice as large as what it is at large bias, where the coupling through Δ becomes inefficient. This factor of 2 is preserved up to eV$|= \Delta$. It is a direct signature that the condensed state in S is composed of pairs. While Giaever tunneling probes the nature and density of excited states in S above the gap, ASJ reflections can probe the condensate itself below the gap. The condensate manifests itself through coherence peaks at the gap edge in Giaever tunneling, and through the enhanced conductance in ASJ reflections below the gap. The gap is identified by the position of the coherence peak in Giaever tunneling, and by the decrease of the enhanced conductance in ASJ reflections (Fig. 5.1).

The general case: Blonder–Tinkham–Klapwijk model

The general case of a contact of arbitrary transparency has been treated by Blonder, Tinkham and Klapwijk (BTK). The contact is modeled as a delta function barrier $V = H\delta(x)$, and is characterized by the dimensionless parameter $Z = (H/\hbar v_F)$. An electron coming from the N side can be scattered in four different ways, each of them having a probability which depends on the bias and on the value of Z : A is the probability that it will undergo an ASJ reflection as a hole backtracking the trajectory of the incoming electron; B is the probability that it will be specularly reflected (normal reflection); C is the probability that it will be transmitted as an electron having a wave vector $k > k_F$; and D the probability that it will be transmitted as an electron having a wave vector $k < k_F$. The sum of the

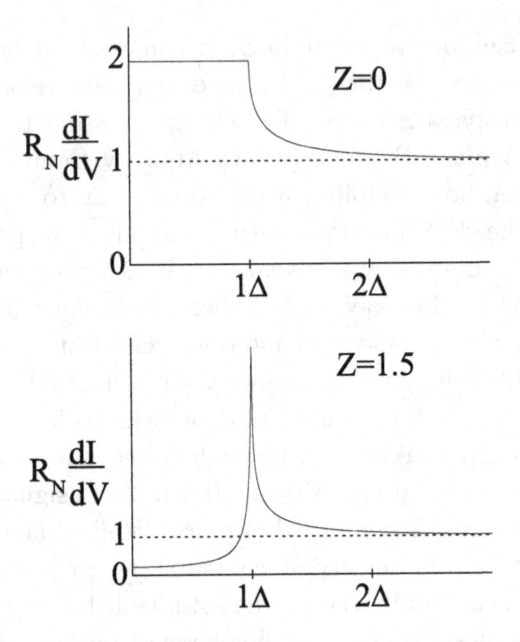

Figure 5.1: Normalized conductances of a transparent S/N contact (no interface barrier) (upper graph) and a non-transparent one (lower graph). The gap edge is marked by a sharp decrease of the conductance in the first case and by a coherence peak in the second one.

probabilities must be equal to 1. Since current is conserved across the interface, one can write it for instance as the current in N:

$$I = J_0 \int_{-\infty}^{+\infty} [1 + A(\epsilon) - B(\epsilon)][f(\epsilon - eV) - f(\epsilon)] d\epsilon \tag{5.13}$$

where f is the Fermi function and J_0 is a conductance taking into account the geometry of the contact. At T=0, the Fermi function is a step function and the conductance is given by:

$$(dI/dV)_S = (dI/dV)_N [1 + A(\epsilon) - B(\epsilon)] \tag{5.14}$$

In the case of a Sharvin contact (Z=0), for $\epsilon < \Delta$, $B \equiv 0$, A=1 and the conductance has twice the normal state value; for $\epsilon > \Delta$, A

is given by the ratio of the hole to electron character of the excitation propagating in S, (v^2/u^2) :

$$A(\epsilon) = [1 - (\epsilon^2 - \Delta^2)^{1/2}/\epsilon]/[1 + (\epsilon^2 - \Delta^2)^{1/2}/\epsilon] \tag{5.15}$$

Over a range of the order of Δ, $A(\epsilon)$ goes back to zero, and the conductance returns to its normal state value.

In the case of a Giaever contact $(Z = \infty) A \equiv 0$, B=1 for $\epsilon < \Delta$ (the conductance is zero) and for $\epsilon > \Delta$ it follows the density of states as discussed above.

In the general intermediate case, the conductance has peaks at energies $| eV |= \Delta$, and a normalized conductance intermediate between 0 and 2 at zero bias.

5.2 Gap symmetry in the cuprates

There is solid experimental evidence that the gap symmetry in the cuprates is different from that in the low temperature superconductors.

5.2.1 Experimental evidence for low lying states

One of the main features of the s-wave BCS gap is the absence of available states below the gap. Besides tunneling, bulk probes also confirm this absence. They include the exponential decrease of the heat capacity and the exponential saturation of the London penetration depth at low temperatures.

Tunneling in the cuprates is complicated by their quasi two-dimensional character. Early tunneling experiments used junctions made on surfaces parallel to the CuO_2 planes (out-of-plane tunneling), which showed structure-less characteristics with a linear increase of the conductance up to large bias. This increase occurs because in the cuprates out-of-plane tunneling is basically inelastic due to their quasi-two-dimensional character. It prevents coherent electronic motion in the direction normal to the CuO_2 planes along which incoming electrons are focused by the tunnel junction. Because of the lack of coherent out-of -plane states, tunneling is only possible through an inelastic process in which the applied bias provides the energy eV necessary to kick up the incoming electron into an in-plane state of that energy. When inelastic tunneling dominates, the theory of superconducting tunneling of the preceding section does not apply, it completely masks the gap structure.

The in-plane average density of states can be accessed by vacuum tunneling from an atomic size tip brought close to a surface parallel to the CuO_2 planes, as done in a Scanning Tunneling Microscope (STM). STM measures the local density of states, which is an angular in-plane average. Tunneling is in fact in-plane and elastic as is allowed by the uncertainty principle due to the small size of the tip. Most experiments of this kind have been performed on vacuum cleaved Bi 2212 samples. Conductance characteristics measured on optimally doped and even more clearly on overdoped samples show a distinct coherence peak which, as we have discussed above, is a clear signature of superconductivity (the case of underdoped samples is more complicated and will be discussed below in the section devoted to pseudo-gap effects). But the conductance at low bias is not flat as in a conventional s-wave superconductor, even far below T_c it shows a rounded or sometimes sharper minimum indicating the presence of available states below the gap identified by the coherence peak, down to the Fermi level (Fig. 5.2).

One could argue that tunneling only probes the sample surface, whose properties might well be different from the bulk particularly because of the short coherence length. However a similar conclusion regarding the presence of low lying states is reached from the London penetration depth linearly increasing temperature dependence at low temperatures. The penetration depth is directly related to the superfluid density, as we have seen in the first two chapters. Its increase at low temperatures measures the loss of superfluid density due to thermal excitations out of the condensate. In an s-wave superconductor, it varies as $[\exp - (\Delta/k_B T)]$, and remains very small as long as $k_B T \ll \Delta$. The linear increase seen in the cuprates shows that the density of states, instead of remaining equal to zero up to the gap, increases linearly with energy starting from the Fermi level.

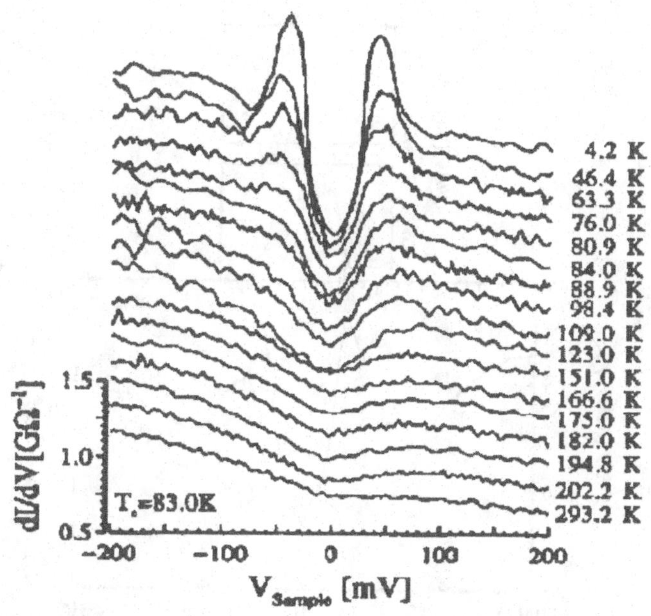

Figure 5.2: STM characteristics of an underdoped Bi 2212 crystal. Coherence peaks disappear at the critical temperature, but a region of depressed conductance persists up to higher temperatures in the same bias range. After C. Renner *et al.*, Phys. Rev. Lett. **80**, 149 (1997).

Another property probing the bulk is the electronic thermal conductivity, which also decreases exponentially at low temperatures in an s-wave superconductor but follows a power law in the cuprates.

5.2.2 Gap anisotropy

The above experiments give an average information on the density of states. They do not indicate where the low energy excitations lie in momentum space. They could be either isotropic, just as in the case of gapless superconductivity induced by magnetic impurities in a conventional superconductor, or anisotropic. We review here two kinds of experiments that demonstrate that the second term of the alternative is the correct one.

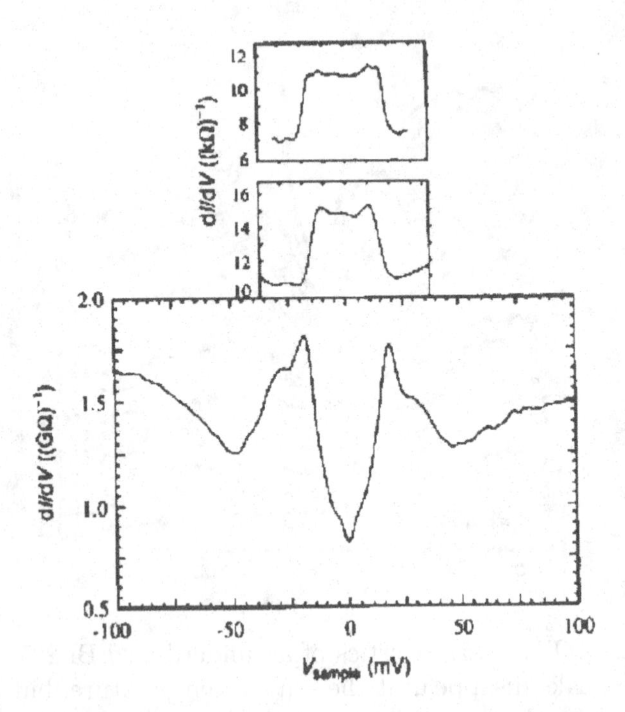

Figure 5.3: In optimally doped YBCO, the gap edge identified by the sharp decrease of the conductance of Sharvin contacts corresponds to the STM coherence peak. See G. Deutscher in further reading for reference.

In-plane ASJ reflections

In-plane ASJ reflections obtained from Sharvin contacts along the a or b axis of CuO_2 planes ([100] or [010] orientation) in optimally doped YBCO single crystal quality samples show characteristics quite similar to those of s-wave superconductors. The conductance is enhanced below an energy of about 20 meV, beyond which it returns to an approximately constant value. This value of 20 meV corresponds to the position of the conductance peak seen by STM in this compound (Fig. 5.3). The conclusion is that in optimally doped YBCO there is indeed one single energy scale, like in a BCS superconductor.

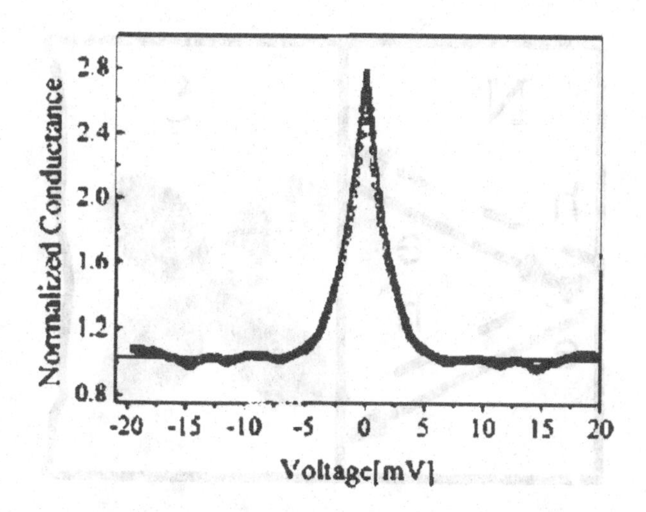

Figure 5.4: Zero Bias Conductance Peak of a Point Contact on a LSCO crystal in the [110] direction. After Dagan *et al.*, cited by G. Deutscher see further reading for reference.

Upon closer examination, the Sharvin contact characteristics do however show some difference from those of s-wave superconductors. From the zero bias conductance enhancement by a factor of about 1.4 (instead of 2 for a perfectly transparent interface), one calculates from the BTK theory a barrier parameter $Z = 0.3$. For a finite Z value, the characteristic should have a peak at the gap edge, where the conductance should be enhanced by more than a factor of 2 above the normal state (high bias) value. Instead, the conductance is almost flat up to the gap edge. This result is inconsistent with an isotropic s-wave order parameter. Additionally, a second, broader conductance peak is observed in the STM characteristic, in a bias range where there is no indication of remaining ASJ reflections.

Zero Bias Conductance Peak: An even more drastic departure from the behavior predicted for an s-wave superconductor is seen in in-plane tunneling at 45° from the principal axis ([110] orientation). There is a large conductance peak at zero bias, called Zero Bias Conductance Peak (ZBCP) (Fig. 5.4). It reveals that for the specific

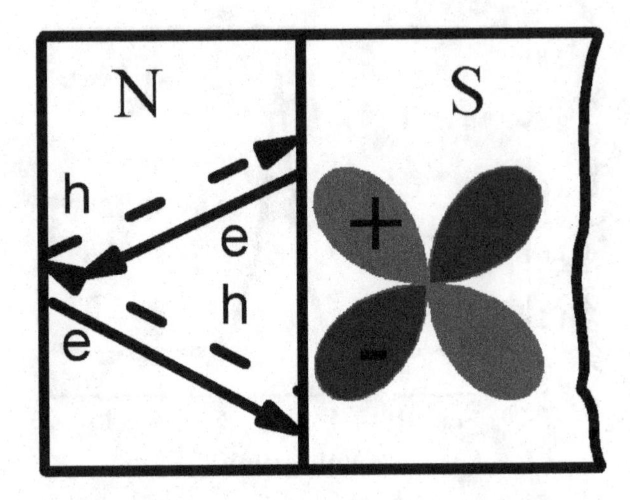

Figure 5.5: The ASJ cycle at an interface having a nodal orientation between a normal metal and a d-wave superconductor. After G. Deutscher, see reference in further reading.

surface orientation there exists a large tunneling zero energy density of states.

This is a surface effect, consistent with a $d_{x^2-y^2}$-wave symmetry gap of the form:

$$\Delta = \Delta_0 \cos(2\theta) \tag{5.16}$$

where θ is the angle with one of the principal axes. This gap changes sign at $\theta = 45^0$. For a [110] tunneling direction the gap is zero, and could be naively assumed that the tunneling density of states should be just that of the normal state. But in fact, this is not what happens. Quasi-particle trajectories near the surface undergo a cycle comprised of successive ASJ reflections from the gap in the bulk, and normal specular reflections at the surface. During one cycle, there are two ASJ reflections from gaps having phases that differ by π, which result in a zero energy state (Fig. 5.5).

All trajectories, not just trajectories perpendicular to the surface, contribute zero energy states. In fact, this will hold true even if scattering at the surface is not specular. Zero energy states are formed

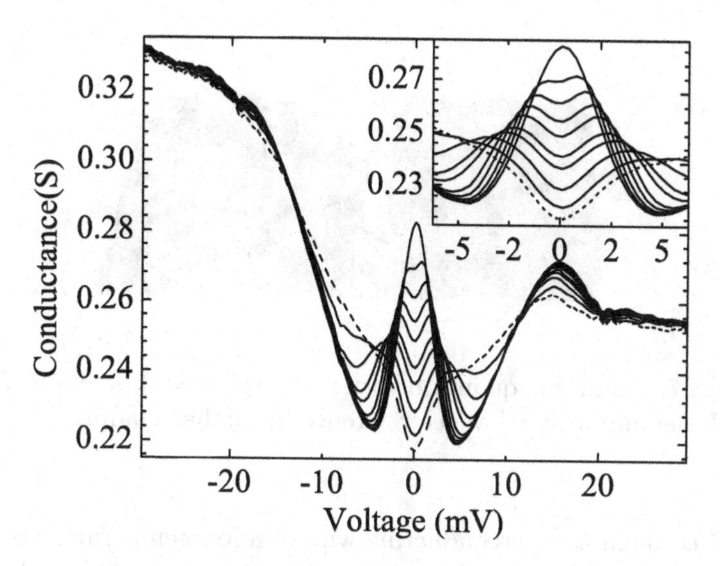

Figure 5.6: Conductance of a (110) oriented In-YBCO tunnel junction in fields of up to 6 Tesla applied parallel to the surface and perpendicular to the CuO_2 planes. Note the Gap-Like-Feature at 16 meV. For a detailed discussion, see G. Deutscher in further reading.

due to the change of sign of the gap, irrespective of its absolute values in the successive ASJ reflections. For the same reason, a ZBCP is observed for all surface orientations except [100] and equivalent, because it is only in that case that no sign reversal of the gap occurs in successive ASJ reflections. Detailed calculations show that the d-wave symmetry gap also explains why there are no strong conductance peaks at the gap edge in the [100] orientation, as noted above (Fig. 5.3). A remnant of the gap structure called "Gap Like Feature" (GLF) is often seen in [110] junctions, which has been attributed to surface roughness (Fig. 5.6).

Interference experiments: The original experiments establishing the d-wave symmetry of the order parameter in the cuprates were performed in multi-junction devices (for a review, see C.C. Tsuei and J. Kirtley in further reading). If two faces of a crystal, one having the [100] orientation and the other the [010] orientation are con-

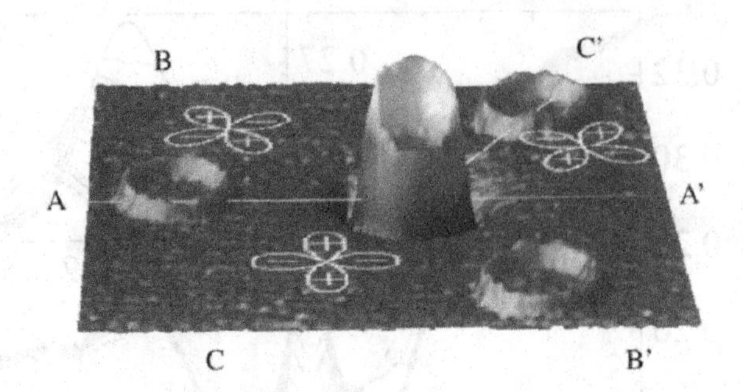

Figure 5.7: Half flux quantum vortex at a tri-crystal junction. After C.C. Tsuei and J. Kirtley, see reference in further reading.

nected through a superconducting wire of a low temperature s-wave superconductor such as Pb, a spontaneous current will circulate in the circuit thus formed because of the phase difference of π at the two contacts. Contrary to an ordinary superconducting loop, whose critical current is maximum when the applied field is zero, here the critical current will be minimal at zero field since π is the maximum phase difference that can be sustained. It will reach its maximum value each time the enclosed flux in the loop is equal to $\Phi_0(n+1/2)$ instead of $n\Phi_0$, where n is an integer.

Spontaneous half flux quanta have been directly detected in tri-crystal junctions, where the three grain boundaries are arranged in such a way that the phase change around a loop surrounding the vertex is equal to π (Fig. 5.7). The nature of grain boundary junctions will be discussed at some length in Chapter 9. They are thought to be due to a depressed order parameter at and near the grain boundary, possibly caused by a lack of oxygen compared to the optimum doping level.

Detection of minority components of the order parameter: Additional components of the order parameter such as s, is or id_{xy} are in principle possible. Imaginary components would remove the node of the order parameter (while a real component would just move it from the 45^0 orientation).

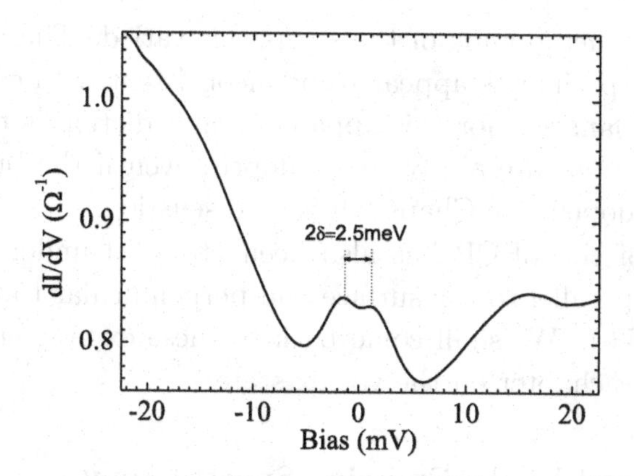

Figure 5.8: Spontaneous split of the Zero Bias Conductance Peak in a [110] oriented In-overdoped YBCO junction.

The ZBCP is a very sensitive tool to study such a possibility, because when the node is removed the ASJ cycle described above, for a certain angular width around the node direction, does not anymore comprise two ASJ reflections from order parameters having a phase difference of π. These cycles do not contribute to the zero energy density of states. The junction conductance is then at a minimum, instead of a maximum at zero bias. The ZBCP appears split.

A systematic study of the ZBCP in junctions prepared on [110] oriented YBCO films has revealed that their shape depends on the doping level. Below optimum doping, the ZBCP is unsplit, in agreement with the presence of a node and consistent with a pure d-wave symmetry. Beyond optimum doping, a spontaneous split develops progressively (Fig. 5.8).

These results can be explained by the progressive apparition of an is or id_{xy} component as doping is increased beyond optimum. As is always the case with tunneling experiments, there is no way to tell whether the minority imaginary component is a bulk or a surface property.

On the other hand, the multi-junction interference experiments mentioned above do not show any measurable deviation from half flux quantum when doping of the samples is varied. These two sets of tunneling experiments appear to be inconsistent with each other. A possible explanation for this apparent contradiction is that grain boundary junctions are always underdoped, even if the bulk of the crystal is overdoped, see Chapter 9 for a discussion.

Splitting of the ZBCP has also been observed under magnetic fields applied parallel to the surface and perpendicular to the CuO_2 planes (Fig. 5.6). We shall come back to these experiments at the end of the next chapter on the vortex state.

Angle Resolved Photo Emission Spectroscopy

In addition to tunneling, another experimental method for the determination of the superconducting gap, called Angle Resolved Photo Emission Spectroscopy, or ARPES (see further reading) has been developed in recent years. In an ARPES experiment, a beam of photons of known energy $h\nu$ is directed at the sample's surface at some angle, and the energy and momentum of emitted electrons are measured. One can then obtain the energy versus momentum of excitations created by the incoming photons.

Principles of ARPES: The analysis must take into account the fact that what is being actually measured is the momentum and energy of the outcoming electron, while what one wants to obtain is the energy versus momentum of the excited electron inside the sample. If the surface is perfect (translation invariant), one can assume that the momentum component parallel to the surface is conserved during the exit of the excited electron. Because of their quasi two-dimensional character, one can in the case of the cuprates make the assumption that the momentum of the excited electron is entirely in-plane. The parallel momentum of the outcoming electron, measured in the experiment, is then equal to that of the excited electron. What was a difficulty in the case of tunneling is an advantage in the case of ARPES.

Denoting by θ the angle that the momentum of the outcoming electron makes with the normal to the surface and by E its energy, one then has the following relation for the momentum of the excited electron:

$$k = \left[(2mE)^{1/2}/\hbar\right]\sin\theta \tag{5.17}$$

According to the Einstein relation, the energy E of the emitted photo-electron is given by:

$$E = E_i(k_i) + h\nu - \Phi \tag{5.18}$$

where $E_i(k_i)$ is the energy of the excited electron inside the sample and Φ is the work function. Since electrons in a metal can have energies up to the Fermi energy E_F, photo-electrons can have energies up to $(E_F + h\nu - \Phi)$. Beyond that edge there will be no photo-electrons. The presence of an edge serves to identify the Fermi level.

The wanted dependence $E_i(k_i)$, usually counted from the Fermi level, can be obtained from the measured quantities E, θ and ν with the above equations.

More generally, what is being measured is the probability $P(E_i, k_i)$ of observing an outcoming electron of energy E and wave vector k originating from an excited electron of energy E_i and wave vector k_i. It is given by:

$$P = \mid M \mid^2 A(E_i, k_i) \tag{5.19}$$

where M is the matrix element for the optical excitation, and $A(E_i, k_i)$ is the spectral function.

If the excited electrons were free electrons, the spectral function would simply be a delta function:

$$A(E_i, k_i) = \delta[E_i(k_i) + h\nu - \Phi - E] \tag{5.20}$$

Angle integration over all k vectors of this δ function then gives the density of states as a function of energy. In an actual solid many body corrections are important. In a conventional Fermi liquid, the life time of excitations diverges at the Fermi level as $(\epsilon - \epsilon_F)^{-2}$, so that on the energy scale relevant for an experiment performed at the finite temperature $T, \epsilon = k_B T$, the excitations look infinitely long lived as the temperature is lowered towards zero. This translates into a maximum amplitude at the Fermi level. In the superconducting state, the piling up of states near the gap is expected to result in an amplitude peak at energy $(E_F - \Delta)$. By following the spectrum P as a function of temperature, one can detect the opening up of the superconducting gap. This measurement can be done as a function of angle.

Two types of information are thus obtained from ARPES.

(i) In the normal state, the Fermi surface can be mapped out and the Fermi velocity determined.
(ii) In the superconducting state, the energy gap is obtained as the range of energy near the Fermi surface where there are no excitations. In practice, most authors present the gap as the shift of the mean point of the leading edge of the photoemission intensity. The energy resolution of detectors has been continuously improved and is now in the range of a few meV, suitable for the measurement of the gap in the HTS. Gap anisotropy can then be directly determined.

ARPES results

The incoming electron beam penetrates only down to a few nanometers below the sample's surface. It is therefore critical that this surface be as "clean" as possible, namely that its structure and electronic properties be as close as possible to those of the bulk. To this aim, samples are cleaved in vacuum. Fortunately some of them, particularly the bismuthates, cleave easily. This is done at low temperature since the loss of oxygen from the surface is always a concern in the cuprates. An alternative route is to grow a cuprate film and measure it without its surface being ever exposed to air.

Results in the normal state of optimally doped and overdoped samples show a hole like Fermi surface (or more exactly a Fermi line since what is being measured are in-plane excitations, the normal to the sample's surface being perpendicular to the CuO_2 planes), see insert of Fig. 5.9. A complete Fermi line is obtained for optimally doped and overdoped samples, establishing that they have a large Fermi surface like ordinary metals do.

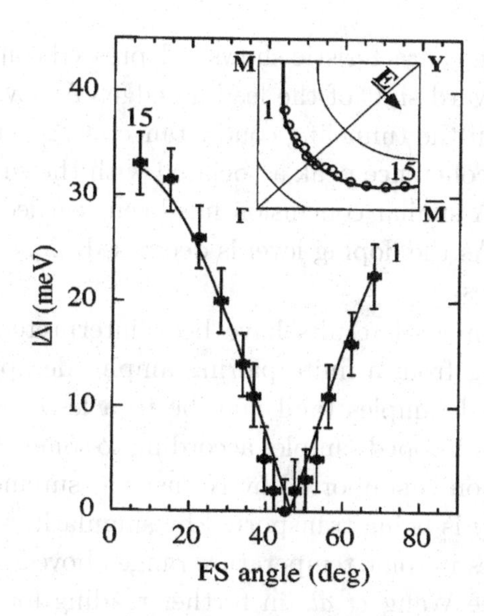

Figure 5.9: Angular dependence of the gap obtained from ARPES data on a Bi 2212 crystal (T_c=87K) by fitting the data to a phenomenologically broadened BCS spectral function. Insert: Fermi surface. After Ding *et al.*, Phys. Rev. B **54**, 9678 (1996).

In the superconducting state ARPES has given a spectacular demonstration of the gap anisotropy in Bi 2212 cleaved crystals. In optimally doped samples, it is consistent with the d-wave symmetry (Fig. 5.9). Gap values estimated from the shift of the leading-edge midpoint (LEM) are smaller than shown in Fig. 5.9. For a number of cuprates (Bi 2201, Bi 2212, Bi 2223, YBCO, LSCO) at optimum doping LEM gap values are of about 0.25 eV$\cdot T_c(K)$.

5.3 Superconducting gap and pseudo-gap

In underdoped samples, spectroscopic measurements (tunneling, ARPES, NMR) have confirmed that the downturn seen in the resistivity (see preceding chapter) below a temperature $T^* > T_c$, is due to a reduced density of states at the Fermi level. Below T^* the spin susceptibility starts to decrease. Up to an energy Δ_p called the pseudo-gap the tunneling characteristic shows a depressed conductance, and ARPES a downward shift of the leading edge. At low temperature, a peak appears in the tunneling conductance at Δ_p, suggesting that it has become a coherence peak associated with the superconducting gap (Fig. 5.2). A similar conclusion has been reached from ARPES measurements. As the doping level is decreased, Δ_p and T^* increase while T_c decreases.

These experimental results have been interpreted by some authors as resulting from a finite pairing amplitude appearing below T^* in underdoped samples (and may be to a lesser extent in optimally doped or overdoped samples according to some measurements). This interpretation is supported by Nernst measurements that indicate that entropy is being transported by singularities that could be similar to vortices in some temperature range above T_c (smaller however than T^*, see Wang *et al.* in further reading for a reference to this work).

A closer look at low temperature tunneling measurements in underdoped samples indicates however that the amplitude of the conductance peak at Δ_p is in general much weaker than predicted by a BCS d-wave density of states, particularly in the most underdoped samples where Δ_p is largest. It is only in overdoped samples that a well developed coherence peak is seen. This suggests that the coherence of the condensed state is actually not fully preserved up to the pseudo-gap energy scale.

ASJ reflections probe as we have seen the condensate rather than excitations above the condensate. They are therefore a powerful tool for the study of coherence. Study of ASJ reflections has shown that in underdoped samples they occur only up to an energy $\Delta_c < \Delta_p$. While Δ_p goes up when doping is reduced, Δ_c follows the same doping dependence as T_c. The two energy scales get closer to each other, and sometimes even become identical, in overdoped samples. While the ratio $(2\Delta_p/k_BT_c)$ becomes very large in underdoped samples, reaching values of more than 10, the ratio $(2\Delta_c/k_BT_c)$ always remains about constant and in the range of 4 to 5, values that are compatible with BCS (Fig. 5.10). In underdoped samples, the absence of a strong coherence peak at Δ_p and that of ASJ reflections

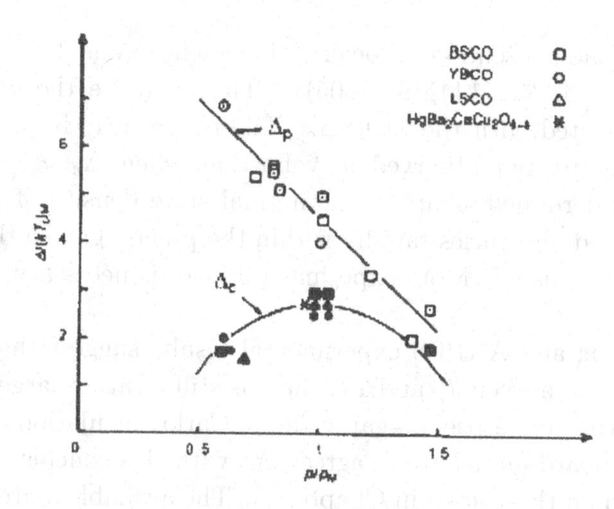

Figure 5.10: Single particle excitation Δ_p and coherence scale Δ_c in various cuprates. The ratio $(2\Delta_c/k_BT_c)$ is compatible with BCS, $(2\Delta_p/k_BT_c)$ is not. For reference, see G. Deutscher in further reading.

up to Δ_p are consistent with each other, and prove that here Δ_p is not the superconducting gap in the BCS sense. Also as indicated by the gap to critical temperature ratio, the energy scale up to which the condensate remains fully coherent is Δ_c rather than Δ_p.

One could then ask why it is that tunneling coherence peaks are not observed at Δ_c? The tunneling density of states seen in YBCO may give at least a partial answer to that question. In optimally doped samples it often shows a double peak structure, a lower one at the energy up to which ASJ reflections are observed (20 meV), and a second one at a higher energy (25 to 30 meV) (Fig. 5.4). It is tempting to associate the lower one with Δ_c, and the higher one with Δ_p. This interpretation is consistent with the fact that the position of the lower peak follows in YBCO the same doping dependence as T_c does, both in the overdoped and in the underdoped regions. Sharp conductance peaks have also been observed in Bi 2212 slightly underdoped high quality single crystals break junctions, below a broader pseudo-gap peak (S.I. Vedeneev and D.K. Maude, Phys. Rev. B **72**, 144519 (2005)). They may be the coherence peaks associated with the scale Δ_c. The reason why in most cases these peaks are not observed may be that when $\Delta_c \ll \Delta_p$ as in strongly underdoped samples, the normal state density of states is much reduced and varies rapidly within the pseudo-gap so that they can easily be missed. More experimental work is necessary to settle this issue.

Tunneling and ARPES experimental results suggest that in the normal state, between T_c and T^*, there is still a rather large density of states within the pseudo-gap. Monte-Carlo calculations of a repulsive Hubbard model are in agreement with this conclusion (more will be said on this model in Chapter 8). The available states within the pseudo-gap allow ASJ reflections to take place. If, as ASJ experiments suggest, the superconducting state is characterized by a different, smaller energy scale than the pseudo-gap, a first guess for the superconducting tunneling conductance would be:

$$(dI/dV)_S = (dI/dV)_N N(\epsilon) \tag{5.21}$$

where $(dI/dV)_N$ is the pseudo-gap normal state density of states, and $N(\epsilon)$ the superconducting d-wave density of states characterized by the scale Δ_c. Monte-Carlo calculations of the density of states in the superconducting state in the repulsive Hubbard model are not yet available to check this conjecture.

5.4 Summary

Experiments testing the occurrence of low lying states together with those enabling a study of the gap anisotropy both in amplitude and in phase are consistent with a dominant d-wave symmetry in hole doped cuprates.

Two issues remain at this stage unsettled.

The first one concerns the existence of a subdominant imaginary component of the order parameter. While experiments probing Andreev–Saint-James surface bound states are consistent with the existence of such a component in overdoped YBCO (at least at the surface), others such as the tri-crystal experiments of Tsuei and Kirtley are not.

The second one concerns the interpretation of the pseudo-gap observed in all hole underdoped cuprates above the critical temperature, and the interpretation of the low temperature gap in these samples. While some authors have interpreted the pseudo-gap as a manifestation of a pre-formed pairing amplitude, and the low temperature gap as a continuation of the pseudo-gap into the condensed state where it becomes the superconducting gap, the absence of strong tunneling coherence peaks and that of ASJ reflections up to that gap still remain to be understood. An alternative would be that there are two distinct energy scales in the condensed state of underdoped cuprates, the larger one being that of the normal state (pseudo-gap) and the smaller one that of the condensed state. The existence of a double peak in the density of states of YBCO is well established and consistent with this view, but more experiments on other cuprates are necessary to check whether it is a general feature of underdoped cuprates.

In this chapter we have only reviewed the gap structure of hole doped cuprates. We have left aside electron doped cuprates whose gap properties are under intense study and still not so clear.

5.5 Further reading

For a recent review on tunneling and Andreev–Saint-James reflections, see G. Deutscher, Rev. Mod. Phys. **77**, 109 (2005).

For a review on interference experiments, see C.C. Tsuei and J. R. Kirtley, Rev. Mod. Phys. **72**, 969 (2000).

For a review on ARPES spectroscopy see A. Damascelli, Z. Hussain and Z.X. Chen, Rev. Mod. Phys. **75**, 473 (2003).

For recent Nernst effect measurements see Yayu Wang, Lu Li and N.P. Ong, cond-mat/0510470.

Chapter 6

Basics on vortices

6.1 Vortices and vortex matter

We have already introduced in Chapter 2 two characteristic fields: the nucleation field H_n, and the thermodynamical critical field H_c. In a type II superconductor, the nucleation field is also the upper critical field H_{c2}. Below H_{c2}, the superconductor is penetrated by flux lines, called vortices, each of them carrying a flux quantum Φ_0. Screening currents surround each vortex, extending up to a radius of order λ, while the superfluid density is suppressed at the center of the vortex, in a region called the vortex core, which extends over a range of order ξ (see Fig. 2.3).

In the presence of a current flowing perpendicular to it, a vortex line is submitted to a Lorentz force:

$$\mathbf{F}_L = \frac{\mathbf{j} \times \Phi_0}{c} \tag{6.1}$$

In a perfect, translation invariant sample, any finite current will induce vortex motion. Current then flows in the normal core (instead of by-passing as it does when the vortex is at rest), inducing dissipation. Vortex motion and dissipation can be avoided by pinning down vortices at defects.

As we show below, vortices in a low T_c superconductor can be considered as rigid entities. Up to H_{c2} they form a lattice which plays an important role in determining the conditions under which vortices can be prevented from moving. The lattice elastic properties allow a single defect to pin down a bundle of vortices of size

λ. In a high T_c–high κ Type II such as the High T_c cuprates, line rigidity is diminished, interaction between the vortices is weaker, the temperature is higher and the lattice can melt below H_{c2}. The phase diagram is then much richer, and has been the object of intense studies in recent years. Because of the different states that can exist — such as vortex lattice, vortex glass, vortex liquid — the object of this sub-field of study has been called "vortex matter". It is a new and exciting field in superconductivity which has obviously important practical implications for high field–high current applications of High T_c cuprates, because a diminished line rigidity and lattice melting can profoundly modify the extent to which pinning centers can prevent vortex motion.

The properties of vortex matter are closely connected with the classification of superconductors that we have established previously, in terms of the condensation energy per coherence volume. We first start with a brief review of some of the main features of the vortex state in conventional Type II superconductors, before proceeding in the next chapter to review some aspects of vortex phases in the High T_c, which are strongly influenced by the small value of this condensation energy.

6.2 The isolated vortex

6.2.1 The line energy

It is convenient to characterize a vortex by its line energy L per unit length. This line energy consists of three terms: the condensation energy lost in the vortex core, L_c; the magnetic energy L_M; and the kinetic energy L_K of the screening currents circulating around the core:

$$L = L_c + L_M + L_K \tag{6.2}$$

We see immediately that:

$$L_c \approx \Delta F.\xi^2 \tag{6.3}$$

where $\Delta F = H_c^2/8\pi$ is the condensation energy per unit volume. Remembering that $H_c = \Phi_0/2\pi\sqrt{2}\lambda\xi$, L_c can be rewritten as:

$$L_c = \frac{1}{2}\left(\frac{\Phi_0}{4\pi\lambda}\right)^2 \tag{6.4}$$

The magnetic term is of the same form, because the magnetic field average value is (Φ_0/λ^2) and its squared value has to be integrated over the vortex magnetic area, of order λ^2.

To evaluate the kinetic energy term, we must integrate the kinetic energy of the superconducting particles over the vortex area:

$$E_K = \int d\mathbf{r}\frac{1}{2}mv_S^2 n_s \tag{6.5}$$

where v_s is the superfluid velocity and n_s the superfluid density. The superfluid velocity and the superfluid current are related by:

$$\mathbf{j} = n_s e \mathbf{v}_s \tag{6.6}$$

From the Maxwell equation:

$$\mathrm{curl}\,\mathbf{h} = \frac{4\pi}{c}\mathbf{j} \tag{6.7}$$

we obtain:

$$E_K = \frac{1}{8\pi}\int d\mathbf{r}\frac{mc^2}{4\pi n_s e^2} \mid rot\mathbf{h} \mid^2 \tag{6.8}$$

Since from the London equation $\mid rot\mathbf{h} \mid$ is of the order of the average field value divided by the London penetration depth, itself equal to $\left(mc^2/4\pi n_s e^2\right)^{1/2}$, E_K has also the same form as E_c. The exact result for the total line energy is:

$$L = \left(\frac{\Phi_0}{4\pi\lambda}\right)^2 \left[\ln\frac{\lambda}{\xi} + 0.11\right] \tag{6.9}$$

6.2.2 Vortex rigidity and pinning

The state of lowest energy for a vortex is to be straight and aligned parallel to the applied field. Any deviation from this orientation would produce a restoring couple $\Phi_0.\mathbf{B}$. And any deviation from a straight line would add to the line length, hence to the effective line energy. In all discussions of the vortex state in conventional Type II superconductors, it is indeed implicitly assumed that vortex lines are basically straight and rigid. Under this assumption, Abrikosov has shown that they form a lattice, and that this lattice persists up to the upper critical field H_{c2}.

Let us evaluate the energy cost of a local deformation in an otherwise straight vortex line. The smallest relevant scale for this deformation is the coherence length and according to Eq. (6.9) its energy cost is given:

$$U_{\text{local}} \approx \left(\frac{\Phi_0}{4\pi\lambda}\right)^2 \xi \tag{6.10}$$

This is an interesting result: the energy cost of the smallest relevant deformation is of the order of U the condensation energy per coherence volume, $U = \Delta F \cdot \xi^3$. At such deformations, the order parameter will be depressed in a "bubble" having a volume of the order of ξ^3 (Fig. 6.1). We have seen that for a conventional superconductor, U is by many orders of magnitude larger than $k_B T_c$. Therefore, the probability that any meaningful deformation of the vortex from a straight line can be thermally excited is vanishingly small. For these superconductors, the assumption made by Abrikosov that vortices are straight and rigid is perfectly justified.

As already said, for practical applications of Type II superconductors it is essential to prevent vortex motion. Any defect that induces a spatial dependence of the vortex line energy will tend to localize it at locations where this energy is at a minimum. To be specific, let us consider the case where small normal (nonsuperconducting) precipitates have been introduced in the superconducting matrix. If a line goes through such a precipitate, its core energy will be reduced, since inside the precipitate the order parameter is in any

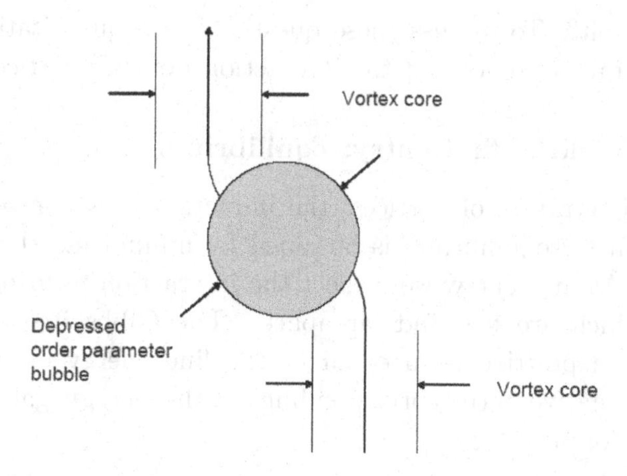

Figure 6.1: A local deformation of a vortex line involves a region of depressed order parameter due to the phase gradients produced by the deformation.

case zero. The line will tend to be "pinned" at such locations. For a precipitate having a size of order ξ, the depth of the potential well it sits in is of the order of the condensation energy per coherence volume.

The characteristic energy scale involved in vortex rigidity and in vortex pinning is thus the same one. In low T_c superconductors it is quite large as we have seen, vortex rigidity and vortex pinning can persist up to H_{c2}.

But we have already seen, for instance from heat capacity measurements presented in Chapter 3, that cuprates are not generally in the regime $U \gg kT_c$. The question of vortex rigidity and vortex pinning needs then to be re-examined. This will be discussed in the next chapter.

6.3 Formation of the vortex lattice

As we increase the applied field, when do vortices start to penetrate inside the superconductor? Why do they form a lattice? Can this lattice melt? To discuss these questions in a quantitative way, we need to take into account the interaction between vortices.

6.3.1 Field of first entry: equilibrium

Upon penetration of vortices, the induction B is non-zero and the condition for equilibrium is obtained by minimizing the Gibbs free energy. At first entry, we neglect the interaction term between vortices, which are few and far apart. The Gibbs free energy then contains a positive term equal to the line energy of the vortices, and a negative term corresponding to the energy gained by field penetration:

$$G = n_L.L - \frac{BH}{4\pi} \tag{6.11}$$

Since each vortex carries a flux quantum:

$$B = n_L \Phi_0 \tag{6.12}$$

where n_L is the vortex density.

There will be a gain in energy when the applied field will be larger than:

$$H_{c1} = \frac{4\pi L}{\Phi_0} \tag{6.13}$$

or:

$$H_{c1} = \frac{\Phi_0}{4\pi\lambda^2} \ln\left(\frac{\lambda}{\xi}\right) \tag{6.14}$$

At equilibrium, the vortex state extends in the range $H_{c1} < H < H_{c2}$.

6.3.2 Field of first entry: Bean–Livingston barrier

Let the applied field be parallel to the surface of the superconductor. Assume that a vortex line has penetrated inside, and that it is located at a distance $x = r$ from the surface, not very far from it on the scale of the penetration depth. We wish to calculate the forces to which this line is submitted. Currents surrounding the line must obey the boundary condition that they have no component normal to the surface. We can ensure this boundary condition by adding an image anti-vortex, located at $x = -r$. This image anti-vortex exerts an attractive force on the vortex. As shown by Bean and Livingston (see further reading for references), this force can prevent the penetration of the vortex in the bulk, up to a certain value of the applied field, H_s. In the range $H_{c1} < H < H_s$, the Meissner state can persist as a metastable state.

We can estimate this field as the field at which the superfluid velocity of the induced Meissner currents reaches the maximum possible value, namely the depairing value:

$$H_s = \frac{4\pi}{c}\lambda n_s e V_c \tag{6.15}$$

Using the relations $V_c = \frac{\Delta}{p_F}$, $\lambda^{-2}=4\pi n_s e^2/mc^2$ and $\xi = \frac{\hbar v_F}{\pi\Delta}$, this can be rewritten as:

$$H_s = \frac{2}{\pi^2}\left(\frac{\Phi_0}{\lambda\xi}\right) \tag{6.16}$$

This field is of the order of the thermodynamical field H_c, $H_s = (2\sqrt{2}/\pi)H_c$. In the Ginzburg–Landau approximation, the four fields H_{c1}, H_s, H_{c2} and H_{c3} have the same temperature dependence. For a long Type II cylinder in an increasing parallel magnetic field, the Meissner state persists up to H_s, the bulk vortex state extends up to H_{c2}, beyond which surface superconductivity persists up to H_{c3}. In decreasing fields, surface superconductivity nucleates at H_{c3}, the vortex state at H_{c2}, and is replaced by the Meissner state at H_{c1}.

6.3.3 Vortex interactions and lattice formation

In our calculation of H_{c1}, we have neglected the fact that several vortices enter actually simultaneously. Since at fields of order H_{c1} the distance between these vortices is of order λ, at which the density of the screening currents is small, their interaction is negligible. Considering a single vortex line to calculate the value of H_{c1} is thus justified. But as the field is increased, vortices are packed more closely and their interaction must be taken into account.

We can estimate this interaction by writing that the force exerted on vortex 2 due to the presence at a distance r of vortex 1 is:

$$\mathbf{F}_{1,2}(r) = \frac{\mathbf{j_1}(r) \times \mathbf{\Phi_0}}{c} \tag{6.17}$$

To calculate the current surrounding vortex 1 at distances $r > \xi$, we can use the London equation:

$$\mathbf{h} + \lambda^2 \operatorname{curlcurl} \mathbf{h} = 0 \tag{6.18}$$

since in this region the order parameter is essentially constant. Near the vortex core, we replace the r.h.s. by $\Phi\delta(r)$. Upon integration up to a radius r, the second term of the l.h.s. transforms into a line integral of $\operatorname{curl} \mathbf{h} = 4\pi j/c$ along a circle of that radius. If $r \gg \lambda$, currents are negligible and:

$$\int \mathbf{h} \cdot d\boldsymbol{\sigma} = \Phi \tag{6.19}$$

Hence, $\Phi = \Phi_0$:

$$\mathbf{h} + \lambda^2 \operatorname{curlcurl}| \mathbf{h} = \Phi_0 \delta(r) \tag{6.20}$$

If we now integrate up to a radius $r \ll \lambda$, the integral $\int \mathbf{h} \cdot d\boldsymbol{\sigma}$ will be negligible compared to Φ_0, leaving:

$$\operatorname{curl} \mathbf{h} = \frac{\Phi_0}{2\pi\lambda^2 r} \tag{6.21}$$

And finally we get for the force exerted by vortex 1 on vortex 2 at distances $\xi \ll r \ll \lambda$:

$$F_{1,2}(r) = \frac{1}{8\pi^2} \frac{\Phi_0^2}{\lambda^2 r} \tag{6.22}$$

We calculate the interaction energy $U_{1,2}$ between the two vortices as the work necessary to bring vortex 2 from infinity to distance r from vortex 1. At larger distances $r \gg \lambda$, the interaction energy falls off exponentially. Neglecting the contribution to the energy integral of the part extending from infinity to λ, we obtain per unit length and per vortex:

$$U_{1,2} = \left(\frac{\Phi_0}{4\pi\lambda}\right)^2 \ln\left(\frac{\lambda}{r_{12}}\right) \tag{6.23}$$

For inter-vortex distances $\xi \ll r_{12} \ll \lambda$, $U_{1,2}$ is of the same order as the line energy, Eq. (6.9). This repulsive interaction is the basis for the formation of the Abrikosov vortex lattice, whose properties have been calculated exactly in the vicinity of the upper critical field. In particular, its elastic constants are known under the assumptions that vortex lines are rigid and that fluctuations of this lattice are negligible. We have already examined under what condition the assumption of line rigidity is valid. We now examine under what condition lattice fluctuations can be neglected.

6.3.4 Lattice deformations

To be specific, let us consider a deformation of the lattice in which a line is brought out off its equilibrium position. For simplicity, we consider a row of vortices, and displace line n by bringing it closer to line $(n+1)$ and further away form line $(n-1)$, by a displacement δr. In a strong Type II, $\kappa \gg 1$, and in fields substantially higher than H_{c1}, the induction B is very close to the applied field H, so that the equilibrium distance between vortices is given by $a = \left(\frac{\Phi_0}{H}\right)^{1/2}$. Neglecting the logarithmic term, the energy cost of the displacement is then, per unit length:

$$\delta U = \left(\frac{\Phi_0}{4\pi\lambda}\right)^2 \left(\frac{\delta r}{a}\right)^2 \tag{6.24}$$

Now, what we envision is of course a local deformation over some finite length of the line. The smallest meaningful deformation has an amplitude $\delta r = \xi$, and must at least extend over a similar scale. Hence, the energy cost of the smallest possible deformation of the lattice is of order:

$$\delta U_{\min} \approx \left(\frac{\Phi_0}{4\pi\lambda}\right)^2 \frac{\xi^3}{a^2} \tag{6.25}$$

We can rewrite this expression as:

$$\delta U_{\min} \approx U \left(\frac{\xi}{a}\right)^2 \tag{6.26}$$

where U is our familiar condensation energy per coherence volume. Within a factor of order π, this expression is equivalent to:

$$\delta U_{\min} \approx U \left(\frac{H}{H_{c2}}\right) \tag{6.27}$$

In a conventional low T_c superconductor, U is by many orders of magnitude larger than $k_B T_c$ (except of course very close to T_c). Hence, as implicitly assumed by Abrikosov, lattice fluctuations will be completely negligible. Vortices are, as we have seen above, straight and rigid, and because of their strong repulsive interaction they form a solid lattice. This, however, will not in general be the case in the cuprates. Consequences, which include lattice melting, will be examined in the next chapter.

Another interesting remark concerns the effect of the applied field on the deformation energy. Starting from low fields, this energy first increases as the lattice becomes denser, which enhances its rigidity. On the other hand, as one approaches H_{c2} where a second order phase transition takes place, U becomes strongly field dependent. It goes down progressively to zero, and the lattice becomes softer against deformation. We therefore have a re-entrant behavior: as the external field is increased, the lattice first becomes more rigid, and then softer. This general result is in practice irrelevant for low T_c superconductors, because U is so much larger than $k_B T_c$. But it is of interest for the cuprates.

6.3.5 The Abrikosov lattice near H_{c2}

We briefly recall here some known exact results for the superfluid density near the upper critical field, as they will be useful in the next chapters. As H_{c2} is approached, the distance between vortices is of order ξ. In the space between vortices, the order parameter does not recover its bulk zero field value. Its mean square value tends linearly to zero:

$$\langle | \Delta(r) |^2 \rangle = \frac{ck_BT}{2eN(0)D} \frac{H_{c2} - H}{2\kappa^2 - 1} \psi^{(1)} \left(\frac{1}{2} + \rho \right) \tag{6.28}$$

where $\rho = (DeH/2\pi ck_BT)$, D being the coefficient of diffusion, $N(0)$ the density of states at the Fermi level and $\psi^{(1)}$ the trigamma function. This expression is valid in the dirty limit. For our purpose, it is more appropriate to express this result in terms of the field dependence of the average superfluid density. From the known relation:

$$n_s = \alpha \frac{m}{\hbar^2} N(0) \xi^2 \mid \Delta \mid^2 \tag{6.29}$$

where α is a number of order unity, we get in the limit $\kappa \gg 1$:

$$n_s(H) = \eta n_s(0) \frac{H_{c2} - H}{H_{c2}} \tag{6.30}$$

where η is a number of order unity.

As for the magnetization, it is given by:

$$-4\pi M = \frac{H_{c2} - H}{(2\kappa^2 - 1)\, \beta_A} \tag{6.31}$$

where β_A=1.16 for the triangular lattice.

In the intermediate field range $H_{c1} \ll H \ll H_{c2}$, an approximate expression for the magnetization is:

$$-4\pi M = \frac{\Phi_0}{8\pi\lambda^2} \ln \left(\frac{H_{c2}}{2(H - H_{c1})} \right) \tag{6.32}$$

6.4 Vortex motion

Consider a slab in the vortex state, with the magnetic field oriented perpendicular to the slab along the z axis, and a current flowing along the y axis. The Lorentz force is then oriented along the x axis. If the vortex moves with velocity v_x, the dissipation is given by:

$$W = E_y.J_y \tag{6.33}$$

with:

$$E_y = \frac{Bv_x}{c} \tag{6.34}$$

The problem is to calculate the velocity v_x in the presence of pinning defects.

Suppose that pinning defects are normal precipitates having a size of order ξ. This is the optimum size for the defect: at smaller sizes it is less effective, at larger ones too much condensation energy is lost for the matrix, at no benefit. If the vortex core intercepts such a precipitate, it sits effectively in a potential well whose depth is given by the condensation energy per coherence volume, our known quantity U, as discussed in Sec. 6.2.2.

This pinning model is appropriate for low T_c superconductors, for which as we know $U \gg k_B T_c$. The potential well is quite deep on the scale of the thermal energy, and the escape probability is small. But the line can still "hop" from pinning site to pinning site at some finite rate, which will set the average line velocity, and therefore the dissipation. Actually, since the vortex lattice is quite rigid, we should not think in terms of motion of isolated lines, but in terms of motion of "flux bundles" , of typical lateral size λ. The interaction between lines is, as we have seen, so strong that one cannot modify much the distance between them. Even if only one line goes through a pinning defect, a whole bundle can be pinned (Fig. 6.2). We do not need a large concentration of defects to pin the entire lattice. Motion will occur through "flux jumps", rather than through hopping of single vortex lines.

In the presence of a current, the energy landscape looks qualitatively like a "washboard". Upon escaping a potential well, the bundle will move preferentially in the direction given by the Lorentz force. We can distinguish two regimes.

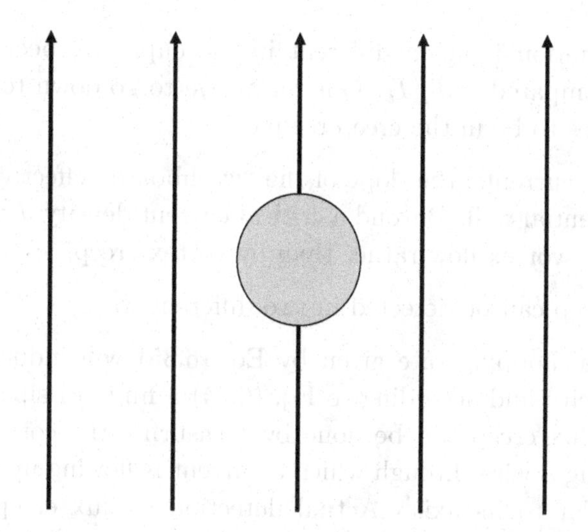

Figure 6.2: In a low T_c, relatively long coherence length superconductor, one defect is capable of pinning a bundle of vortices due to line rigidity and strong vortex interactions that prevent lattice fluctuations.

(i) At very low currents, the effective depth of the potential well is unaffected by the current. The hoping rate is then given by:

$$\tau^{-1} = R_0 \exp\left(-\frac{F_0}{kT}\right) \tag{6.35}$$

where R_0 is some attempt frequency, and F_0 is the condensation energy per coherence volume if the applied field is near H_{c2}, and the condensation energy per volume is λ^3 if the applied field is of order H_{c1}. The field variation of F_0 reflects the number of vortices per bundle. In any case, for a low T_c superconductor, $F_0 \gg k_B T_c$ as we have seen. The hopping rate is very small, dissipation is low. This is the "creep" regime. Notice that it is always helpful to operate the superconductor at the lowest possible temperature, no matter what T_c is, since dissipation goes down with temperature even far below the critical temperature. This is the reason why large coils are often operated below the λ point of liquid Helium.

The situation is again different in the cuprates, because F_0 is not large compared to $k_B T_c$. One may have to go down to very low temperatures to be in the creep regime.

(ii) At large currents, the slope of the "washboard" effectively wipes out the potential well. Beyond a critical current density $J_{c\infty}$, vortex motion is by vortex flow rather than by vortex creep.

Flux creep can be detected in two different ways:

(i) The final hopping rate given by Eq. (6.35) will induce a finite vortex velocity and according to Eq. (6.34) a finite dissipation. Detection of flux creep can be done by measuring the corresponding voltage along a wire through which a current is flowing under a field perpendicular to its axis. Actual detection of flux creep will depend upon the sensitivity of the instrumentation used. A commonly retained criterion for determining the value of the critical current, which in fact corresponds to a certain flux line velocity, is 1μV/cm. At that level of dissipation the line velocity is of 10^{-2} cm/sec.
(ii) Flux pinning results in an inhomogeneous induction. Under increasing fields applied parallel to the surface of a cylinder, the induction gradient is given by $(4\pi/c)J_c$, where J_c is an induction dependent critical current density. Flux creep will result in a slow motion of flux lines from high density to low density regions, eventually resulting in an homogeneous induction in the sample. By measuring the sample magnetization over long periods of time, very slow flux line motion can be detected. This second method is relevant if one wishes to test the suitability of a given superconducting wire for the fabrication of superconducting coils used as permanent magnets (such as for NMR applications).

6.5 Probing surface currents in d-wave superconductors

Surface currents in the vortex state are closely related to the magnetization, which is in general irreversible due to defects (including the surface Bean–Livingston barrier). It turns out that the d-wave symmetry of the order parameter opens up the possibility to probe surface currents by tunneling experiments. This is because, when the field is applied parallel to the surface and perpendicular to the CuO_2 planes, the zero energy surface states that are at the origin of the ZBCP seen on [110] surfaces have their energy shifted by a Doppler effect by a term $\mathbf{p}_F.\mathbf{v}_s$ where $\mathbf{p}_F$ is the momentum of the incoming tunneling particle and $\mathbf{v}_s$ the superfluid velocity of the superconducting particles flowing near the surface of the CuO_2 planes.

Surface currents are quite different depending on the way the applied field has been reached. There are four kinds of surface currents:

(i) Equilibrium currents j_{eq}. Equilibrium situation is best approached by cooling the sample in a field H. The local field in the sample reaches the value of the bulk induction B within a distance from the surface corresponding approximately to the position of the first vortex line. That distance is about the same as the inter-vortex distance in the bulk, $(\Phi_0/B)^{1/2}$.

$$j_{\text{eq}} = \frac{c(H-B)}{4\pi}\left(\frac{B}{\Phi_0}\right)^{1/2} \tag{6.36}$$

or using the approximate expression for the magnetization for $H \gg H_{c1}$:

$$j_{\text{eq}} = \frac{c(\Phi_0 B)^{1/2}}{8\pi\lambda^2}\ln\left(\frac{H_{c2}}{2H}\right) \tag{6.37}$$

The field dependence of the surface equilibrium current is dominated by $B^{1/2}$, which for a very high κ value is the same as $H^{1/2}$ since at the fields of interest the induction and the applied field are very close to each other. Equilibrium surface currents occur over a depth $(\Phi_0/B)^{1/2}$.

(ii) In increasing fields, screening currents can increase linearly up to the field H_s because of the Bean–Livingston barrier that opposes vortex entry. These currents occur over a depth λ.
(iii) In decreasing fields, there is no barrier against vortex exit. Induction in the sample is almost equal to the applied field. Currents related to the existence of the surface are small compared to the two previous situations.

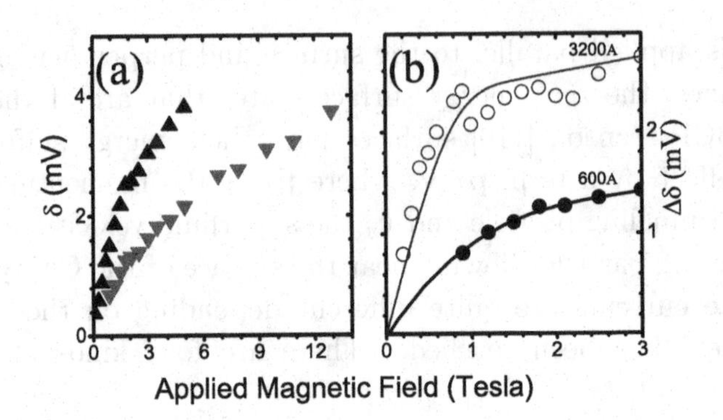

Figure 6.3: Field induced splitting of the Zero Bias Conductance Peak in [110] oriented In-YBCO junctions. (a) Split in increasing (upper triangles) and decreasing fields (lower triangles). (b) Difference between the splits in increasing and decreasing fields for two YBCO film thickness, which reflects the effect of the Bean–Livingston barrier. Note the saturation above 1 Tesla. After R. Beck *et al.* Phys. Rev. B **69**, 144506 (2004).

(iv) In addition to the above, there is a fourth kind of current due to vortex pinning in the bulk, called the Bean current. In the Bean critical state model, under increasing fields induction decreases linearly in the sample with a slope set by the critical current density. That current density reverses sign upon field reversal. It extends up to the depth were the local field reaches zero, or up to the center of the sample, whichever occurs first. Usually, in the presence of strong fields, the critical current is much weaker than the depairing current, so that the external field is screened over a length much larger than λ.

Equilibrium currents, Bean–Livingston currents and Bean currents can all contribute to the field splitting of the ZBCP. In agreement with the above classification, the ZBCP splitting is largest in increasing fields (δ_{up}) and smallest in decreasing ones (δ_{down}), see Fig. 6.3(a). The difference ($\delta_{\mathrm{up}} - \delta_{\mathrm{down}}$) saturates at the field H_s, which can thus be identified, Fig. 6.3(b). The surprise in the data shown in Fig. 6.3(a) is the large split seen in decreasing fields. One possible interpretation is that it is due to a field induced imaginary component of the order parameter having the d_{xy} symmetry.

6.6 Summary

In low temperature superconductors the large condensation energy per coherence volume guarantees vortex line rigidity, efficient vortex pinning and the persistence of the Abrikosov lattice up to the upper critical field.

Because of their reduced condensation energy, these three properties are not expected to be verified in high T_c superconductors.

A d-wave symmetry of the order parameter allows surface currents in the vortex state, strongly dependent on the magnetic history, to be probed by tunneling experiments thanks to the Doppler shift of Andreev–Saint-James bound states.

6.7 Further reading

For an in-depth description of the vortex state in low temperature superconductors, see D. Saint-James, G. Sarma and E.J. Thomas, "Type II Superconductivity", Pergamon Press 1970.

Chapter 7

Cuprate superconductors under strong fields

Soon after the discovery of the High T_c cuprates, it became apparent that their behavior under applied magnetic fields was very different from that of Low Temperature Superconductors (LTS). Noticeably, it was discovered that their magnetization became reversible above a field $H_{irr}(T)$ that could be much smaller than H_{c2}. For instance in the Bi 2212 compound having a critical temperature of 92K, H_{irr}(77K) is only a few 100G, while H_{c2} is expected to be of the order of a few 10 Tesla. In the compound YBCO, having the same critical temperature, H_{irr}(77K) can reach several T, much higher than in Bi 2212, but substantially lower than H_{c2}. The electrical resistance becomes finite above $H_{irr}(T)$, since a reversible magnetization means that there is no pinning; these findings thus shed some serious doubts on the practicality of using HTS under strong fields, for instance, to make superconducting magnets operating at the temperature of liquid nitrogen. This problem was (and still is) particularly acute for the Bi cuprates.

A low value of $H_{irr}(T)$ may not be a fundamental property, as it can be due to the absence of appropriate pinning centers. A perfect crystal of a LTS such as Nb will show a very low value of the irreversibility field. But as pinning centers are introduced, it can rise up to H_{c2}. Considerable effort was invested in introducing all sorts of pinning defects in the cuprates, for instance, creating by irradiation columnar defects that could in principle pin vortices parallel to them very effectively. But these efforts resulted only in modest improvements of the irreversibility field, particularly in the Bi compounds, where it remained far below the values required for making coils operating at liquid nitrogen or thereabout.

The empirical conclusion must therefore be that the low irreversibility field has a fundamental origin. This was actually predicted already in the early 90's, based on the same classification that determines the respective importance of fluctuation effects in the LTS and HTS (for references see G. Deutscher in further reading).

(i) If the condensation energy per coherence volume is much larger than $k_B T_c$, vortices can be considered as rigid straight lines, as we have shown in the previous chapter. They form the well known Abrikosov lattice whose thermal fluctuations can be neglected. This is the case of Type II metals and alloys, where this energy can be of the order of several eV. The Abrikosov lattice persists practically up to H_{c2}.
(ii) If the condensation energy per coherence volume is of the order of $k_B T_c$ or smaller, vortices can easily deform on the scale of the coherence length (Fig. 6.1). At high field and temperature vortices form a liquid rather than a solid well before the applied field reaches H_{c2}. This explains the low value of H_{irr}, since the absence of a lattice facilitates the motion of individual vortices. A limiting case is that where vortex loops can form freely: one can then expect that all coherence is lost. Before this limiting situation is reached, other interesting effects can occur.

The distinction between these two cases, which we have on several occasions already introduced in preceding chapters, is therefore of practical as well as of fundamental importance. The idea developed in this chapter is that the condensation energy per coherence volume, which can be determined from heat capacity, is an excellent guide for the practical use of new superconductors.

7.1 Vortex lattice melting

We shall now try to make the above considerations more quantitative. Consider a case where a magnetic field $H_{c1} < H < H_{c2}$ is applied at low temperatures, k_BT being small compared to any other relevant energy scale. We then have an ordered Abrikosov lattice. We heat up the sample, keeping the applied field constant, and ask up to what temperature this lattice will persist, into what kind of vortex state it will then transform and how new vortex states will affect the critical current.

There are basically two approaches to this problem. One of them consists in treating it as the melting of a solid occurring at some temperature $T_m(H)$. An estimate of $T_m(H)$ can be obtained by using the Lindemann criterion according to which melting occurs when the amplitude of lattice vibrations is a certain fraction c_L of the lattice spacing. The second approach emphasizes the importance of fluctuations of the order parameter.

7.1.1 Lattice melting — Lindemann criterion

Since we are considering applied fields well in excess of H_{c1}, the induction B in the sample is not very different from the applied field. The vortex density n_L can then be approximated by:

$$n_L = \left(\frac{H}{\Phi_0}\right) \tag{7.1}$$

and the vortex lattice spacing by:

$$a = \left(\frac{\Phi_0}{H}\right)^{\frac{1}{2}} \tag{7.2}$$

From Eq. (6.24), the cost in energy of a lateral displacement δr from equilibrium of a vortex line, over a length l, is given by:

$$U(l) = \left(\frac{\Phi_0}{4\pi\lambda}\right)^2 \left(\frac{\delta r}{a}\right)^2 l \tag{7.3}$$

At melting, we set $(\delta r/a) = c_L$.

A rough guess is that the relevant length scale l for the deformation of the lattice leading to its melting is the lattice parameter itself.

We now write that such deformation has a significant probability to occur if its energy cost is of the order of k_BT:

$$k_BT \approx \left(\frac{\Phi_0}{4\pi\lambda}\right)^2 c_L^2 \left(\frac{\Phi_0}{H}\right)^{1/2} \tag{7.4}$$

The melting field is then given by:

$$H_m \approx \frac{\Phi_0}{(k_BT)^2}\left(\frac{\Phi_0}{4\pi\lambda}\right)^4 c_L^4 \tag{7.5}$$

which is valid for an isotropic superconductor. In the case of the cuprates, it is necessary to take the anisotropy into account. The result for a field applied parallel to the *c*-axis of cuprate superconductors is then (for references see D. Nelson in further reading):

$$H_m \approx \frac{\Phi_0}{(k_BT)^2}\left(\frac{\Phi_0}{4\pi\lambda_{ab}}\right)^4 c_L^4\Gamma^2 \ln\kappa \tag{7.6}$$

where λ_{ab} is the London penetration depth characterizing screening currents in the *ab* planes, Γ is the anisotropy ratio (λ_{ab}/λ_c) and $\kappa = (\lambda_{ab}/\xi_{ab})$. This expression has some interesting properties:

(i) near T_c, the melting field varies as $(T_c - T)^2$.
(ii) H_m decreases linearly as the square of the anisotropy factor.
(iii) H_m varies as the square of the superfluid density.

We note that this expression does not contain the coherence length.

It is convenient to write the melting field in the form:

$$H_m = H_m(0)\left(\frac{T_c - T}{T}\right)^2 \tag{7.7}$$

where H_m gives the scale of the melting field. To get an order of magnitude, we take λ_{ab}=$2.2{\cdot}10^{-5}$cm, Γ=1/30, T_c=100K, numbers that are close to those for BSCCO (2212). For c_L=0.1, this gives for $H_m(0)$ a value of the order of 0.1Tesla. The prediction for this cuprate is therefore that melting occurs well before the upper critical field is reached at temperatures of the order of T_c. This is in good agreement with experiment, see below for experimental details. For YBCO, the penetration depth is somewhat smaller and mostly the anisotropy factor is much smaller, 5 instead of 30. The predicted order of magnitude of $H_m(0)$ is 10 to 100 Tesla, also in rough agreement with experiment. However, as we shall see below, the temperature dependence does not fit Eq. (7.7).

7.1.2 Melting and order parameter fluctuation effects

As noted, the coherence length is not involved in the above expression of the melting field. This is because fluctuations of the order parameter have not been taken into account.

Limit of validity of the Lindemann expression

An obvious flaw in the above argument is that it does not set a lower limit to the displacement of vortex lines. This lower bound should be the coherence length, since displacements of a vortex line by less than this scale are meaningless. Since according to the Lindemann criterion melting occurs for a fixed ratio c_L of the displacement to the lattice spacing, for our expression of the melting field to be valid it is required that $c_L a \gg \xi$. This is rewritten as the condition:

$$H_m \ll c_L^2 H_{c2}$$

which becomes with the Lindemann expression for H_m:

$$\left(\frac{U}{k_B T}\right) \Gamma c_L \ll 1$$

within a numerical factor of order unity. This condition is easily met in the high anisotropy oxides such as Bi 2212, for which the reduced condensation energy per coherence volume $u = U/k_B T$ is small. We expect the Lindemann criterion to apply to these cuprates. And indeed, as we shall see when we review experimental results, it does. But for the cuprates belonging to the YBCO family, for which the condensation energy is not small compared to $k_B T$, this condition will not be met (except very close to the critical temperature where U goes to zero). In that case, we have no reason to expect the Lindemann expression for the melting field to apply.

Melting triggered by order parameter fluctuations

We now consider the situation where at the calculated Lindemann field the vortex displacement is not much larger than the coherence length, as is expected for the YBCO family. The next field value where a transition can occur is then that where thermodynamical fluctuations of the order parameter will by themselves produce local lateral displacements of the vortex line of the order of the coherence length. We can write that this is the field H_f where the cost of a displacement of order ξ is $k_B T$. From Eq. (6.24):

$$\delta U(\xi) = \left(\frac{\Phi_0}{4\pi\lambda}\right)^2 \left(\frac{\xi}{a}\right)^2 \xi \tag{7.8}$$

With $\delta U(\xi) = k_B T$, $\left(\frac{\xi}{a}\right)^2 = \left(\frac{H}{H_{c2}}\right)$, we obtain within a coefficient of order unity:

$$H_f = H_{c2}.u(T) \tag{7.9}$$

where $u(T) = (U(T)/k_B T)$.

When that field is reached, melting can take place immediately since the vortex displacement is larger than $c_L a$. This expression is of course only meaningful if $u(T) <1$. In order to facilitate comparison with experiment, we can rewrite Eq. (7.11) in the form:

$$H_f = H_{c2}(0) \cdot u(0) \cdot \left(\frac{T_c - T}{T}\right)^{3/2} \tag{7.10}$$

where $u(0) = U(0)/k_B T_c$ is the reduced condensation energy per coherence volume at T=0.

H_f has two interesting properties:

(i) its scale is set by the upper critical field times the reduced condensation energy per unit volume.
(ii) if varies as $(T_c - T)^{3/2}$, rather than as $(T_c - T)^2$ as does the classical Lindemann field.

In Eq. (7.12) we have used mean field critical exponents, which is questionable since we are considering strong fluctuation effects. But this approximation should not affect the scale of H_f.

3D XY melting

Another approach, also based on the assumption that thermodynamical fluctuations of the order parameter are at the origin of melting, has been proposed for the case where the superconducting transition can be convincingly described as a 3D XY transition (for references see Junod *et al.* in further reading). The only known such case is that of YBCO at optimum doping, as reviewed in Chapter 3. It is then assumed that melting occurs for a fixed value of the ratio (ξ/a), just as H_{c2} does. The temperature dependence of the coherence length is then governed by the 3D critical exponent ν=2/3, which gives for the temperature dependence of both the melting field and H_{c2} a power law $(T_c - T)^{4/3}$.

7.1.3 Loss of line tension

After melting of the vortex lattice has occurred, vortices are free to wander around. Collective pinning is lost, but vortices can still be pinned individually. Up to what field is such individual pinning possible?

Near T_c large thermodynamical fluctuations of the order parameter occur in the critical region. These fluctuations can generate vortex loops, even in the absence of any applied field. Their size diverges as T_c is approached. In the presence of an applied field, these loops can "mess up" the Abrikosov lattice, by combining with field generated vortices. Field induced vortices then lose their individuality, or in other terms their line tension.

What we have been trying to describe in words is actually a very complex situation. To study vortex loops created by fluctuations in the critical region is already a difficult problem. When a magnetic field is applied, it becomes even more difficult: what we are asking is how the applied field modifies the divergence of the loop size that characterizes the critical regime. Analytical solutions are not available, therefore one must resort to numerical calculations to investigate this regime. These simulations do indicate that the applied field does modify this divergence, which can now occur beyond the zero field critical region, in a region of the field-temperature plane that lies within the liquid phase. Besides the melting line, the phase diagram now has a line that separates the liquid phase in two regions: immediately after melting, this liquid contains only finite size loops; beyond the new line, the loop size is the system size (for references see Nguyen and Sudbo in further reading).

Let us now try to see how an applied field can affect the vortex loop size. The energy necessary to create a loop cannot be smaller than the core line energy times the diameter of the loop:

$$U_{\text{loop}} \geq 0.1 \left(\frac{\Phi_0}{4\pi\lambda} \right)^2 2\pi R \tag{7.11}$$

where R is the radius of the loop. The typical radius $\langle R \rangle$ of loops generated by fluctuations is given by:

$$k_B T \approx 0.1 \left(\frac{\Phi_0}{4\pi\lambda} \right)^2 2\pi \langle R \rangle \tag{7.12}$$

In zero applied field, the typical loop size diverges at T_c as $(T_c - T)$ (numerical simulations give an exponent of 1.45 rather than 1), and goes to zero as T goes to zero. It is inversely proportional to the superfluid density. An applied field will reduce the superfluid density. This will increase the vortex loop size. It will go to zero, and therefore the loop will diverge, as H_{c2} is approached. We can try a dependence $(H_{c2} - H)^{-\zeta}$ near H_{c2}. In the mean field approximation, ζ=1.

We can distinguish three cases:

(a) at low enough temperatures, $T < T(\xi)$, where $< R >$ as given by Eq. (7.12) is smaller than ξ, there are very few loops

(b) at intermediate temperatures $T(\xi) < T < T(a)$, where $< R >$ is comprised between ξ and a, vortex loops are smaller than the inter-vortex distance of the Abrikosov lattice. In that case the vortex phase will not be very much affected by the vortex loops. But if the lattice is already molten, or if it forms a disordered glass, the condition $\langle R \rangle > \xi$ may be sufficient for the fluctuation induced loops to "mess up" the liquid phase. By combining with field induced vortices, they may result in the loss of individuality of these lines, or in other terms of line tension

(c) if $T > T(a)$, vortex loops will indeed "mess-up" the vortex lattice and *a fortiori* a glassy or molten phase. In this region vortices lose in any case their identity and therefore their line tension.

Calculations in regimes (b) and (c) require numerical simulations. A simplified model for the loss of line tension is presented in the next chapter in the framework of a discussion of an upper bound for H_{irr}.

7.1.4 Effect of disorder on vortex phase transitions

We have so far not taken into account the presence of defects in the crystal and their influence on lattice melting and other manifestations of fluctuation effects. Line rigidity is affected by these fluctuations, which allows vortices to adjust their position in order to benefit from the energy reduction when they pass through pinning defects. One may then wonder if the Abrikosov lattice will actually persist up to the melting field. This question is relevant at temperatures that are not too close to T_c, since in the vicinity of the transition pinning becomes very weak (because the condensation energy goes to zero). If the Abrikosov lattice does not persist up to the melting field, it may transform into a disordered phase before melting occurs. There are clear experimental indications, reviewed in the next section, that such a solid to solid transition does occur at low temperatures prior to melting.

7.2 Experiments on vortex phase transitions

7.2.1 Experimental methods

A variety of methods have revealed the existence of numerous vortex phase transitions in the cuprates. These methods include magnetization, heat capacity, resistivity and expansivity measurements. Coherence in the vortex phase has also been studied by performing transformer type experiments.

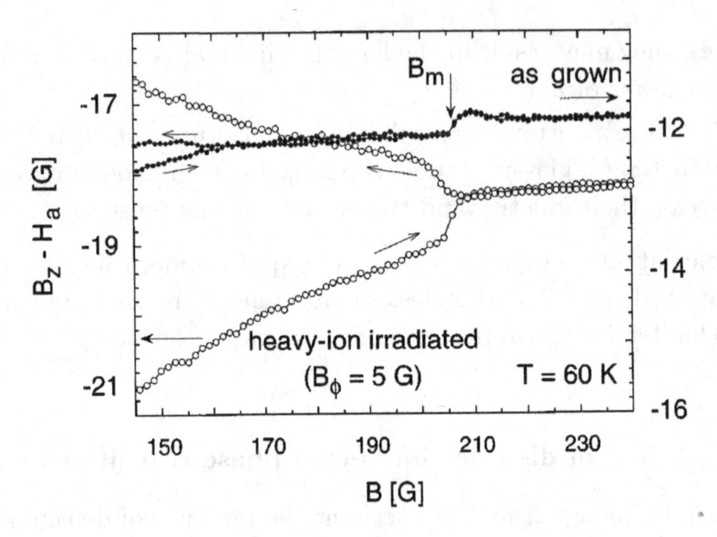

Figure 7.1: For an "as grown" Bi 2212 crystal, the melting field B_m detected by a step in the magnetization is higher than the irreversibility field. Upon irradiation, the latter reaches B_m. After Khaykovitch *et al.*, Phys. Rev. B **56**, R517 (1997).

Observation of a rearrangement of the flux lines by a muon spin rotation experiment, interpreted as the melting transition, was first reported by Lee, Zimmermann, Keller *et al.* (Phys. Rev. Lett, 71, 3862 (1993)), in Bi 2212 single crystals. The first order nature of this transition was put directly in evidence by magnetization measurements performed on similar samples, which showed a finite step at a field $B_m(T)$ (Fig. 7.1). A finite magnetization step identifies a transition as being of first order. Its amplitude does correspond to an increase in entropy of about 0.5 k_B per vortex, which is consistent with melting.

Another signature of a first order transition is a specific heat peak, which was indeed observed on YBCO single crystals at the field where a step in the magnetization takes place. The area under the peak and the amplitude of the magnetization step are consistent with each other (Fig. 7.2).

At the same field-temperature combination, a step down in the resistivity is observed as the temperature is lowered, where it goes from a small but finite value to zero. This can be understood as

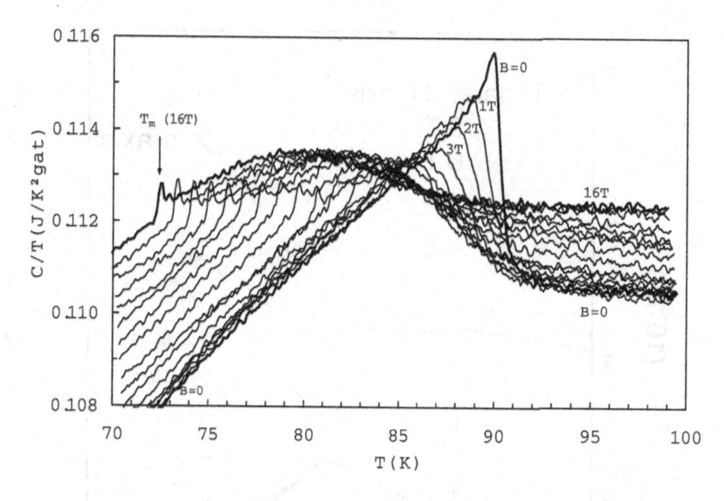

Figure 7.2: The heat capacity of an optimally doped YBCO crystal shows that a first order transition occurs at a temperature $T_m(H)$, up to H=16T. After Revaz *et al.*, Phys. Rev. B **58**, 11153 (1998).

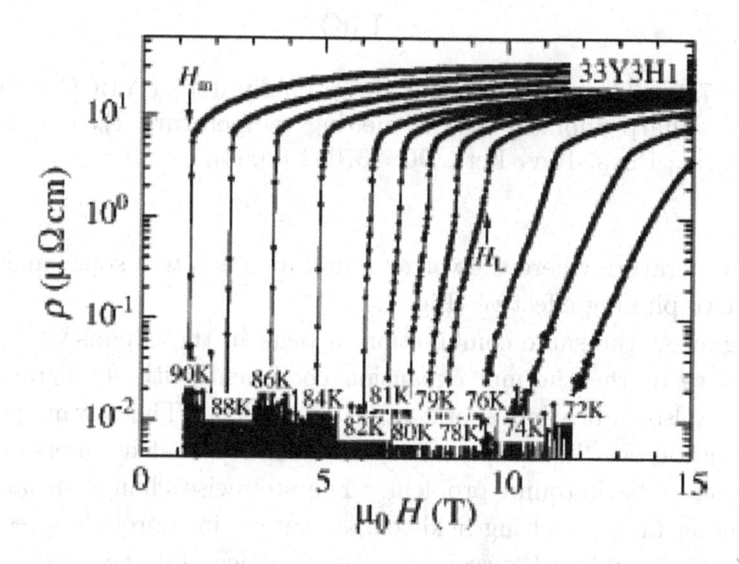

Figure 7.3: The resistivity of an optimal doped YBCO sample shows a sharp step down at the melting field $H_m(T)$. At low temperatures, the transition becomes continuous. After Shibata *et al.*, Phys. Rev. B **66**, 214518 (2002).

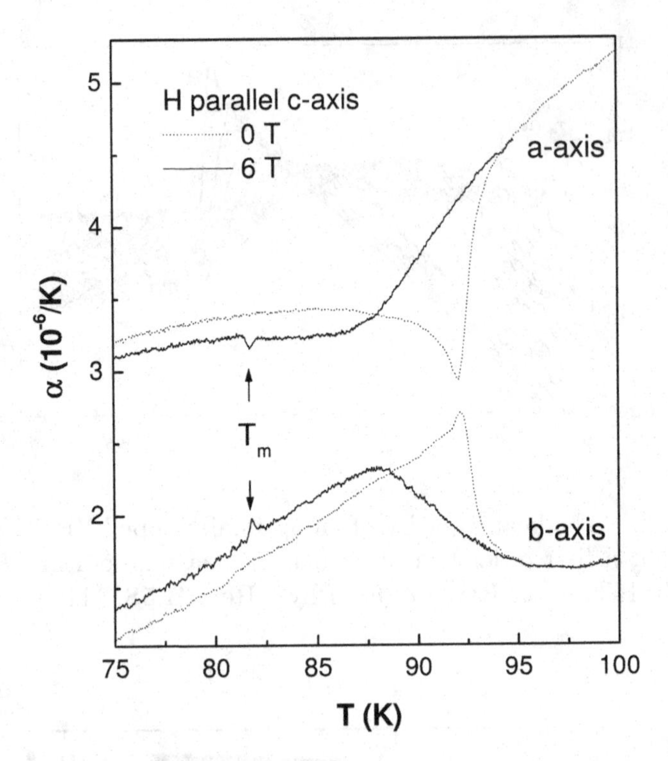

Figure 7.4: The expansivity of an optimally doped YBCO sample shows a sharp anomaly at the melting temperature T_m(H). After Lorz *et al.*, Phys. Rev. Lett. **90**, 237002 (2003).

the temperature where the vortex liquid freezes into a solid, making collective pinning effective (Fig. 7.3).

Again at the same combination, a peak in the expansivity (the derivative of the thermal expansion coefficient with temperature) is seen, also indicating a first order transition. The advantage of this method over heat capacity measurements is that it suffers much less from a background problem. The step-wise change in lattice parameter at the melting field-temperature is interpreted as resulting from a coupling between the vortex lattice and the crystal lattice: the change in density of the vortex phase at melting induces a change in the lattice parameter through the pressure dependence of T_c (Fig. 7.4).

7.2.2 Vortex phase transitions in Bi 2212 crystals

As shown in Fig. 7.1, precise magnetization measurements on high quality Bi 2212 single crystals have shown a small but sharp step occurring at a field $B_m(T)$. In a very clean crystal, $H_{\rm irr}(T) < B_m(T)$. But a few defects are sufficient to raise $H_{\rm irr}(T)$ up to $B_m(T)$. At that dose, and for the fields used here, the concentration of defects created by irradiation is such that only a small fraction of the vortices are pinned, showing that the vortex phase below the first order transition is a solid. Moreover, neutron diffraction experiments have shown that this solid is ordered. It can then be identified with the Abrikosov lattice. As to the vortex phase above the first order transition, it cannot be a solid since vortices are not pinned collectively. The immediate interpretation is that it is a liquid where vortices can still be pinned individually, but not collectively. Alternatively, it could be a gas meaning that vortices lose their line tension at the transition. In a highly anisotropic superconductor, vortices are described as a stack of pancakes. In a solid to liquid transition, that stack retains a line tension. In a solid to gas transition, sometimes called a decoupling transition, each pancake dissociates itself from its neighbors in the adjoining layers.

Further experiments using Josephson plasma frequency measurements have measured the average lateral displacement of the vortices along the stack, called the wandering length r_W. Oscillations of the phase across a Josephson junction involve also charge oscillations. A resonance occurs when the two corresponding energy terms are equal. Vortex wandering lowers the strength of the Josephson coupling, and thus modifies the resonance frequency. The wandering length can thus be determined. In a clean crystal of Bi 2212, it has been shown that r_W remains finite at the first order transition. It was therefore concluded that it is a melting and not a decoupling transition. A close examination of the data shows in fact that for very low fields, where the transition occurs close to T_c, r_W displays a tendency to diverge near the transition. A divergence of the wandering length must be the result of that of vortex loops. This is expected in the critical regime, discussed above. It appears that in that case line tension is lost at the transition (Fig. 7.5).

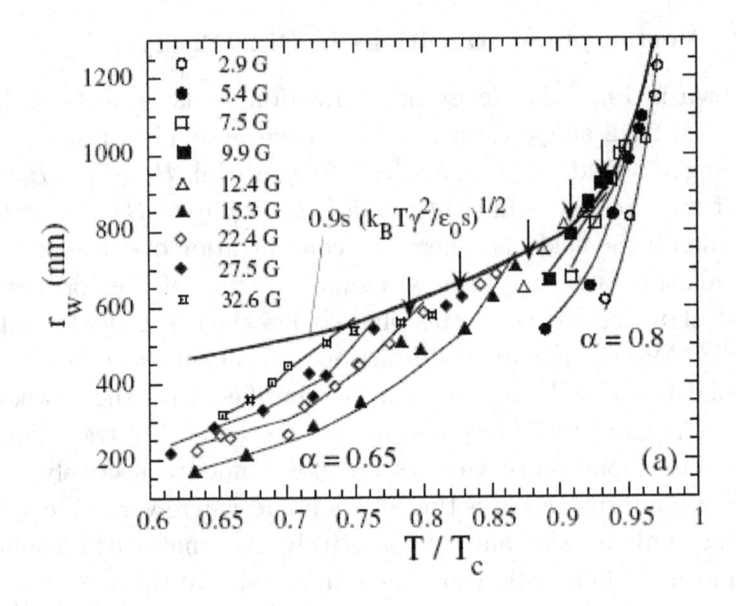

Figure 7.5: The vortex wandering length $r_w(T)$ shows a kink at the melting temperature. It diverges near T_c. After Colson *et al.*, Phys. Rev. Lett. **90**, 137002 (2003).

One may conclude that, except very near T_c, the observed transition is indeed a melting one. The melting field follows a law of the form $H_m = H_m(0) \cdot \left(\frac{T_c - T}{T_c}\right)^\alpha$. The value of the exponent α falls in the range 1.5 to 2. The value of $H_m(0)$, of the order of 1000G, as well as that of the exponent are in reasonable agreement with the predictions based on the Lindemann criterion. This is in line with theoretical expectations since the melting field is much smaller than $c_L^2 H_{c2}$.

The melting field increases with oxygen doping, a dependence also consistent with theory since overdoped samples are known to have a reduced anisotropy.

At low temperatures magnetization measurements as a function of field show a second peak at a field B_{sp} (Fig. 7.6). At that field pinning therefore *increases*. The vortex phase at fields $B > B_{sp}$

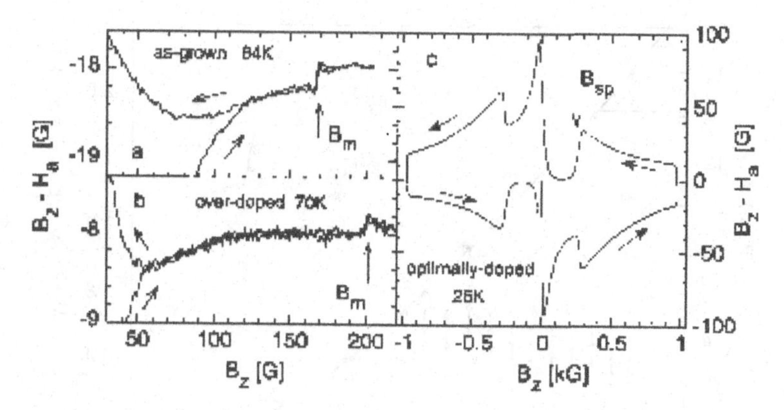

Figure 7.6: At low temperatures, a second peak appears in the magnetization of a Bi 2212 crystal at a field B_{sp}(right panel). At high temperatures, the melting field increases upon overdoping (left panel). After Khaykovitch *et al.*, Phys. Rev. Lett. **76**, 2555 (1998).

must therefore be a solid, not a liquid. This solid phase is a disordered one. It is presumably made more favorable by the reduced line tension as the field is increased. Its extent in the $B-T$ phase diagram increases with the concentration of defects, by a lowering of the field B_{sp}. The lines $B_{sp}(T)$ and $B_m(T)$ connect at a certain field-temperature, possibly a tri-critical point (see below the phase diagram for YBCO). Interestingly enough, the value of the melting field is not modified by the increase in defect concentration, which shows that it is indeed a thermodynamical transition. At large defect density, the melting transition ceases to be a well defined first order transition and becomes a continuous one.

7.2.3 Vortex phase transitions in YBCO crystals

in YBCO single crystals, heat capacity measurements have clearly detected a first order phase transition at a field $B_m(T) < H_{c2}(T)$ (Fig. 7.2). This phase transition manifests itself as a sharp peak, with the area under the peak ΔS being proportional to the applied field (Fig. 7.7).

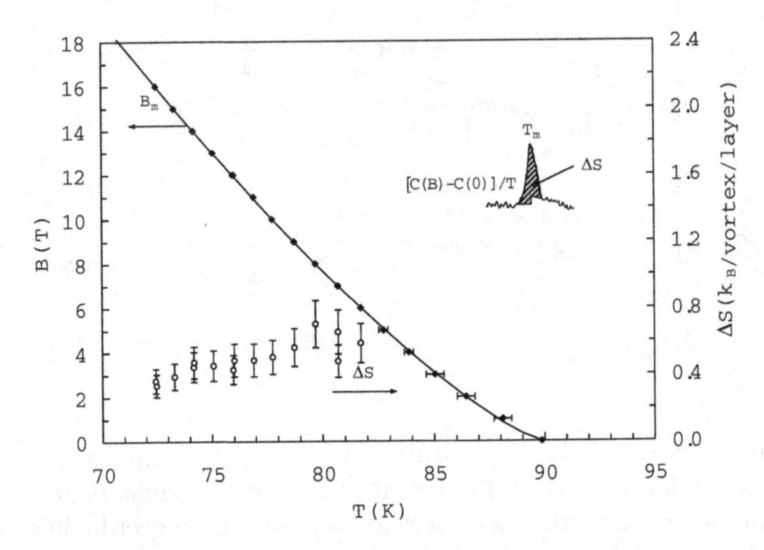

Figure 7.7: The entropy per vortex involved in lattice first order melting (below 82K in this YBCO sample) does not vary with temperature. After Revaz *et al.*, Phys. Rev. B **58**, 11153 (1998).

In other words, the entropy involved in the transition is proportional to the number of vortices, as expected for a melting transition. In addition, the heat capacity peak is coincident with a step in the magnetization, whose magnitude follows the Clausius–Clapeyron relation $\delta S = \left(\frac{dH}{dT}\right)\delta M$. For the best samples, this behavior is observed up to very high fields in excess of 20T. This transition has been attributed to melting of the vortex lattice.

The melting line in YBCO follows a power law temperature dependence $H_m = H_m(0)\left(\frac{T_c-T}{T_c}\right)^{\alpha}$ with $H_m(0)$ somewhat larger than 100T, and α in the range 1.3–1.5. Neither the pre-factor nor the exponent value are in agreement with the Lindemann criterion. This is to be expected, since in the case of YBCO at the Lindemann field the condition $H_m < c_L^2 H_{c2}$ is not met.

The power law exponent measured in YBCO is in agreement with both the predictions of the 3D XY model, and with the expression of the melting field obtained under the assumption that melting occurs

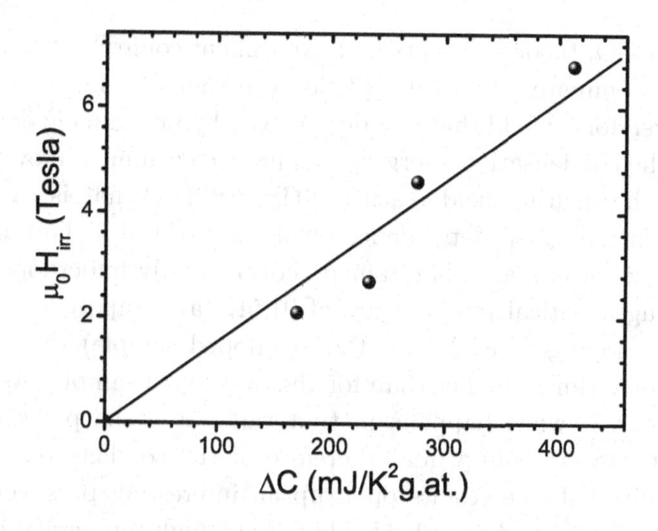

Figure 7.8: The irreversibility field in YBCO samples (here at 15K below T_c) is a linear function of the mean field heat capacity jump at T_c. Oxygen doping levels are O_7 (T_c=86.4K),$O_{6.95}$ (T_c=91.7K), $O_{6.90}$ (T_c=92.7K) and $O_{6.86}$ (T_c=90.3K), in order of decreasing values of ΔC (see Fig. 3.7).

when fluctuations of the order parameter can generate vortex lines distortions of the order of ξ, Eq. (7.12). The fitted value of $H_m(0)$ being of the order of $H_{c2}(0)$, one may conclude that the condensation energy per coherence volume in YBCO at $T = 0$ is of the same order as $k_B T_c$. This is in agreement with estimates of the width of the critical region in YBCO.

Doping dependence of the melting field

Another consequence of Eq. (7.12) is that $H_m(0)$ should vary as $U(0)$, or in other terms as the mean field heat capacity jump at the transition, ΔC. This jump has been measured for a series of $YBa_2Cu_3O_{7-\delta}$ samples having different degrees of doping around optimum doping (Fig. 3.7). As seen in Fig. 7.8, the melting-irreversibility field is indeed proportional to the heat capacity jump. The fast variation of ΔC and $H_m(0)$ with doping makes this agreement convincing.

In YBCO, theory predicts and experiment confirms that the factor that dominates the melting field, and therefore in practice also the irreversibility field that can be achieved by introducing defects, is indeed the condensation energy per coherence volume. In overdoped YBCO, the melting field reaches 30T at 60K. What is extremely striking is the sharp dependence on doping of the highest field up to which zero resistance is retained. For a slightly underdoped sample having a critical temperature of 91.5K (as compared to 93K at optimum doping and 92K for the overdoped sample) that field at 60K is three times smaller than for the overdoped sample. As shown in Chapter 3, what happens in that range is a sharp decrease of the heat capacity jump, and therefore of the condensation energy with doping. These results open up an interesting perspective for the use of overdoped samples for high field–high temperature applications. The critical temperature is not a very good indicator of that potential. A much better one is the amplitude of the mean field jump in the heat capacity (for reference see G. Deutscher in further reading).

Disorder effects

At lower temperatures, the phase diagram comprises not only the Abrikosov and the liquid phase, but up to two additional ones. One of them is the vortex glass phase, as also found in Bi 2212 crystals. Its limits are the glass temperature Tg where the electrical resistance vanishes upon cooling, and the temperature corresponding to the field where the second peak in the magnetization occurs. In an intermediate range of temperatures, an additional transition occurs at a field H_L within the liquid phase. It separates a liquid phase where some pinning subsists and a liquid phase without pinning. The field H_L might be that where line tension vanishes.

Oxygen deficiency (under-doping) increases disorder. As a result, the range of the vortex glass phase is enhanced, at the expense of the Abrikosov phase. Even a relatively slight under-doping is sufficient to suppress the first order melting transition at all temperatures.

7.3 Summary

In the cuprates, the Abrikosov lattice melts before the upper critical field is reached. Melting occurs through different mechanisms in the Bismuthates and in YBCO.

In the Bismuthates, lattice melting is well accounted for by the Lindemann criterion. It applies when the condensation energy per coherence volume is much smaller than $k_B T_c$. In that case, the melting field is very small compared to the upper critical field.

In YBCO, melting is governed by fluctuations of the order parameter producing vortex deformation on the scale of the coherence length. This mode of melting applies when the condensation energy per coherence volume is of the order of $k_B T_c$. The melting field remains of the order of the upper critical field, but is considerably reduced near T_c.

In the presence of disorder, a solid to solid transition from the Abrikosov lattice into a solid disordered phase may occur before melting.

7.4 Further reading

For an introduction to the concept connecting the heat capacity jump at T_c and the potential for high critical currents in high fields, see G. Deutscher in "Coherence in High Temperature Superconductors", Eds. G. Deutscher and A. Revcolevschi, World Scientific 1996.

For an introduction to lattice melting see D. Nelson in "The vortex State", Eds. N. Bontemps, Y. Bruynseraede, G. Deutscher and A. Kapitulnik, NATO ASI series Vol. 418, Kluwer Academic Press 1994.

For a discussion of the relation between lattice melting and the 3DXY model of the superconducting phase transition, see A. Junod *et al.*, Physica B **280**, 214 (2000).

For a discussion of topological phase fluctuations, see A. K. Nguyen and A. Sudbo, Phys. Rev. B **60**, 15307 (1999) and Z. Tesanovic, Phys. Rev. B **59,** 6449 (1999).

Chapter 8

From fundamentals to applications

8.1 The need for high critical temperatures and fields

Since the discovery of superconductivity, it has been a constant aim of researchers to raise the critical temperature, the ultimate goal being of course to reach room temperature. Transport of electricity without losses over long distances, ultra-low loss motors, generators and transformers, would be some obvious applications of room temperature superconductivity. In many of these applications, superconductors would simply replace copper wire. Indeed, for most of them the cost of cooling rules out the use of low temperature superconductors. The discovery of the cuprates, which have practical superconducting properties at liquid nitrogen temperature, has been an important step towards the introduction of superconductors in these applications, but it is not yet clear that it is a decisive one. Thus the quest for still higher temperature superconductors remains an important one. Some theoretical considerations regarding this issue will be found in the next section.

A related but distinct issue, that of the highest magnetic fields that can be sustained by the superconducting state, has received somewhat less attention, even though it is not less important. One may recall that the first application of superconductivity that was considered by Kammerlingh Onnes was a coil generating a high magnetic field. He reasoned that in the absence of losses in the coil, there was in principle no limit to the field that could be achieved. Alas, the only superconducting materials available at the time were Type

I, whose upper critical field is the thermodynamical critical field H_c, which is typically of the order of a few hundred gauss. Superconductivity in the coil is quenched when that small field value is reached.

The idea that the generation of high fields would be an important application of superconductivity was finally implemented some fifty years later with the discovery and understanding of Type II superconductivity. Like pure Nb, the widely used NbTi alloy has a modest critical temperature of only 9K, but NbTi coils can generate magnetic fields of up to 8 to 10T, while pure Nb can only sustain a field of 0.2T. Today, the most widely spread application of superconductivity is in Magnetic Resonance Imaging (MRI). In MRI machines, a field of a few Tesla is generated by a large NbTi coil cooled in a bath of liquid Helium at 4.2K. Note that here the superconducting wire does not replace copper. The generation of fields of even a few Tesla in volumes of the order of 1 m^3 is simply impractical using Cu wire, due to the huge electrical losses in such a coil. For the same reason, only superconducting coils are used nowadays in large scale magnets for high energy physics research and plasma confinement.

Another important difference between large scale magnets and other applications such as cables is that here the operating temperature is not of great importance. For MRI applications for instance, whether the coil is operated at liquid Helium or liquid Nitrogen temperature is of little consequence. The reason is that due to the small surface to volume ratio (say compared to that of a cable), the cooling costs are not a primary concern. Modern MRI machines need to be refilled with liquid Helium only every few months. Of course, operation at liquid Nitrogen would be preferred, but this is only a secondary consideration (except for remote locations where liquid Helium is not available).

But why do we need higher fields than are now available, and if we do, can HTS help us achieve them? One area where there is an immediate need for higher fields is NMR research magnets, because each increase in field improves NMR analytical capabilities. Another possible area is in future particle accelerators. But possibly the most important application of higher fields will be for fusion machines. At the moment the LTS magnets designed for the ITER project give a field of 12T. Higher fields are desirable to improve plasma confinement, and probably cannot be reached with LTS. Other applications of higher fields might be in Superconducting Magnetic Energy Storage (SMES) devices.

The link between higher T_c and higher critical fields is through the coherence length. We recall that in the clean BCS limit, it is given at $T = 0$ as:

$$\xi = 0.18 \frac{\hbar v_F}{k_B T_c} \tag{8.1}$$

and that the upper critical field is given by:

$$H_{c2} = \frac{\Phi_0}{2\pi\xi^2} \tag{8.2}$$

The highest achievable field in the clean limit varies as the square of the critical temperature (further enhancement of H_{c2} by reducing the mean free path is not very meaningful for HTS since the coherence length is already very short). Therefore higher T_c's are clearly favorable for reaching higher critical fields at low temperatures. At intermediate temperatures, however, as we have seen in previous chapters, flux pinning may vanish well before H_{c2} is reached, so that a closer inspection is necessary to establish to what extent a higher T_c is really useful. This discussion is the object of Sec. 8.3.

Magnets for plasma confinement and superconducting cables are two extreme examples of applications for which a higher upper critical field at low temperature and a high critical temperature are respectively the desired characteristics to be achieved. There are also applications such as motors and generators where one needs a mix of both features. Coils capable of generating fields of a few Tesla at temperatures not too far below that of liquid Nitrogen are necessary to make superconductivity attractive from an economic standpoint and permit its widespread use. It is for such applications that the distinction between upper critical field and irreversibility field is crucial. The question of the ultimate value of H_{irr} that can be obtained by the introduction of appropriate pinning defects is discussed at the end of Sec. 8.3.

8.2 High critical temperatures

From the standpoint of the microscopic theory of superconductivity, the question of the ultimate critical temperature that can be achieved is a most difficult one. A general theory taking into account all possible interactions that can result in a superconducting state is still not available, and may never be. Our aim here is to put in perspective some of the better known theories, and most importantly to introduce the idea of a maximum practical temperature up to which useful superconducting properties can be achieved, irrespective of the microscopic mechanism leading to the condensed state.

8.2.1 BCS theory

The original BCS theory applies well to metals and alloys where the electron-phonon mechanism dominates over the direct electron-electron interaction. The electron-phonon interaction is characterized by the parameter $\lambda = N(0)V$ where $N(0)$ is the normal state density of states at the Fermi level and V the interaction potential, and the electron-electron interaction by the parameter μ^*. For large Fermi surface metals, μ^* is much reduced from the bare electron-electron interaction by screening effects. In the weak coupling limit $\lambda < 1$, the critical temperature is given by:

$$k_B T_c = \hbar\omega_D \exp - \frac{1}{\lambda - \mu^*} \tag{8.3}$$

where ω_D is the cut-off Debye frequency of the phonon spectrum. The effective Coulomb repulsion is given by:

$$\mu^* = \frac{\mu}{1 + \mu \ln \frac{E_F}{\hbar\omega_D}} \tag{8.4}$$

For broad band metals, Morel and Anderson have shown that $\mu^* \lesssim 0.1$. The Jellium model gives $\lambda \lesssim 0.5$. Taking λ=0.4 and a Debye temperature of 300K gives $T_c \approx$10K. The highest critical temperature amongst the elements of the periodic table is 9.2K for Nb. Trying to use the BCS expression as a guide for enhanced T_c values quickly reveals that the parameters that enter into it are not independent. A larger density of states in a simple band structure will be associated with a narrow band, hence with reduced Fermi energy and a larger μ^*. A larger electron-phonon interaction will lead to a softening of the phonon spectrum hence to a lower value of some average phonon frequency.

Still within the BCS theory, there are three ways that can lead to T_c values higher than those of the elements. There are two means by which one can increase the density of states without losing screening. One of them is by having two bands participating in the conduction, a narrow d-band and a broad s-band, the narrow band giving the high $N(0)$ and the broad one the small μ^*. This is the case of the A15 compounds like Nb_3Sn, which reach critical temperatures of more than 20K. The other is by reducing the dimensionality, and making use of the van Hove singularity, see below in Sec. 8.2.3. The second method is to use compounds comprising light elements, in order to get high frequency phonon modes, in the hope that they will couple well to the electron gas. This is presumably the case for the compound MgB_2, in which B atoms form layers with high frequency vibration modes. Its critical temperature of 40K is the highest reached by an electron-phonon superconductor. It is also the only superconductor where there is a clear evidence for two distinct gaps in two distinct bands, a favorable factor because it enhances the phase space for interaction.

It is believed that the T_c of MgB_2 is about the maximum critical temperature that can be obtained by the electron-phonon mechanism in three-dimensional superconductors.

8.2.2 The McMillan strong coupling extension of the BCS theory

The BCS expression is valid when $E_F \gg \hbar\omega_D \gg k_BT_c$, which is the case for most superconducting elements and alloys in the weak coupling limit. For instance, in Al E_F=5eV, $\hbar\omega_D$=$3{\cdot}10^{-2}$eV, k_BT_c=$2{\cdot}10^{-4}$eV. However, in some elements such as Pb and alloys such as PbBi, the ratio $\hbar\omega_D/k_BT_c$ is about 10, and some corrections are necessary. The McMillan expression reads:

$$T_c = \frac{\Theta_{\log}}{1.45} \exp\left(- \frac{1+\lambda}{\lambda - \mu^*(1 + 0.5\lambda/(1+\lambda))} \right) \tag{8.5}$$

where:

$$\lambda = 2 \int_0^\infty \alpha^2(\omega) F(\omega) \frac{d\omega}{\omega} \tag{8.6}$$

and:

$$\ln(\omega_{\log} = k_B \Theta_{\log}/\hbar) = \frac{2}{\lambda} \int_0^\infty \frac{\ln \omega}{\omega} \alpha^2(\omega) F(\omega) d\omega \tag{8.7}$$

These expressions are valid for λ of the order of unity. $F(\omega)$ denotes the phonon spectrum, and $\alpha(\omega)$ the frequency dependent electron-phonon interaction. The phonon density of states is available from neutron scattering experiments and the product $\alpha^2(\omega)F(\omega)$ has been determined from tunneling experiments. Values of λ larger than 2 have been obtained for PbBi alloys. For still much larger values, it has been shown theoretically that $T_c \propto \lambda^{1/2}$. Thus in principle the electron-phonon interaction can lead to very high critical temperatures. The main objection to this possibility is that under such a strong interaction the lattice would become unstable. One way (in fact, may be the only way) to circumvent this difficulty is to imagine a structure composed of different elements, say a stack of ABAB... planes where conduction and strong coupling superconductivity take place in the A planes while the B planes, bound to the A planes by some strong ionic bonds, help hold the structure together. This is exactly the structure of $La_{2-x}Sr_xCuO_4$, which of course does not prove that this cuprate is a strong coupling superconductor.

8.2.3 Density of states effects

To complete this section dealing with the BCS theory and its extensions, we include two approaches that emphasize the role of peaks of the density of states to reach a high T_c value. They both involve a reduced dimensionality. DOS peaks resulting from a quasi-1D character of the A15 compounds seen as composed of three sets of orthogonal Nb chains weakly coupled together were originally proposed by Labbe and Friedel. (For references see Labbe and Friedel in further reading.)

The Van Hove High T_c scenario

It was demonstrated rigorously by Van Hove that in a two-dimensional lattice, the density of states presents a logarithmic divergence at some energy whose value depends on the details of the band structure. In a nearest neighbor tight binding model for a square lattice, the singularity is at mid-band, so that the Fermi level is at the singularity at half filling. This model combines a high density of states at the Fermi level and a broad band, exactly what is required to get a high T_c. Taking for the density of states:

$$N(\epsilon) = N_1 \ln \left| \frac{D}{\epsilon} \right| \tag{8.8}$$

where the energy ϵ is counted from the Fermi energy, N_1 is some fraction of the density of states far from the singularity and D is the width of the singularity, Labbe and Bok showed that the BCS expression for the critical temperature is:

$$T_c = D \exp - \frac{1}{\lambda^{1/2}} \tag{8.9}$$

Compared to the regular BCS expression obtained under the assumption that the normal state density of states is slowly varying (in fact constant on the scale of the phonon energies) near the Fermi level, this expression presents two important modifications. First, the width of the singularity replaces some phonon energy. Second, the coupling parameter appears as a square root. The two effects are favorable. D is an electronic energy scale, and a high T_c is obtained even for weak coupling. It was also shown that the effective Coulomb repulsion is reduced by a logarithmic factor $\ln \frac{D}{\hbar\omega_D}$. This model gives easily T_c values of the order of 100K.

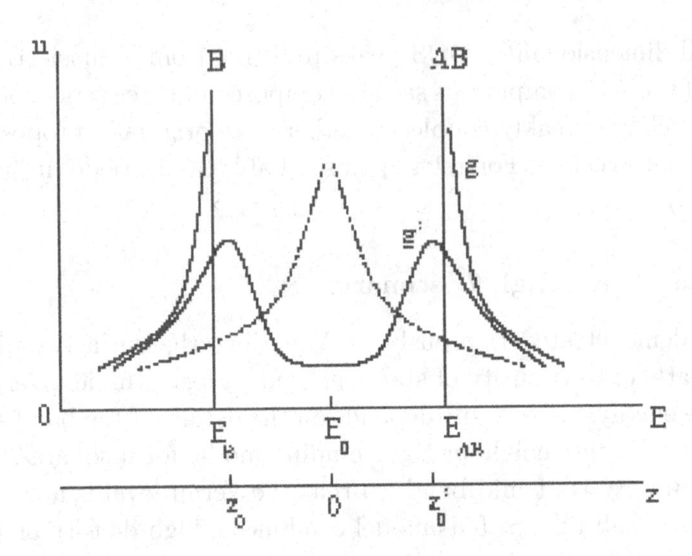

Figure 8.1: Density of states of a pristine (g) and doped (g') band anti-ferromagnet. In the doped case the gap is replaced by a pseudo-gap with a finite density of states at the Fermi level. After Friedel and Kohmoto, Eur. Phys. B **30**, 427 (2002).

DOS peaks in the vicinity of an anti-ferromagnetic state

An insulating anti-ferromagnetic state is in fact known to occur at half filling in CuO_2 planes such as in La_2CuO_4. Therefore the Van Hove scenario does not apply directly to these materials, at least in the simple version above. A central question is that of the nature of the AF state. It can be a Mott insulator, which is at the moment the majority opinion, but it can also be a band insulator resulting from the doubling of the size of the unit cell due to the AF order. Opposite portions of the Fermi surface, being nearly parallel, give strong nesting effects and a quasi-1D character to the split bands. Hence the band structure at half filling consists of two bands that replace the original broad band that one would have without AF order, with peaks of the DOS at the edges (Fig. 8.1). Upon doping (hole or electron), the Fermi level can be made to pass through one of these peaks, resulting in a maximum of T_c at some doping level.

However, doping also introduces magnetic disorder, replacing the AF by a short range AF. Before superconductivity appears, the gap in the DOS becomes a pseudo-gap, with states within the gap (Fig. 8.1). The Neel temperature diminishes and the DOS peaks become broader and weaker. At strong doping, all remnants of the AF order vanish, the original broad band that one would have had without AF order is re-established, and superconductivity is also quenched. This model, proposed by Friedel and Kohmoto, explains qualitatively the basic features of the cuprates phase diagram. One can note that the DOS peaks are of electronic origin and in that sense an expression resembling that for the Van Hove scenario can be expected with a pre-factor being an electronic energy scale rather than a phonon one.

8.2.4 The BCS to Bose–Einstein strong coupling cross-over

The fundamental approximation of the BCS theory is that the gap is much smaller than the Fermi energy, not that it is much smaller than the Debye frequency. As we have seen above, this last limitation has been dealt with by McMillan, still within the BCS theory. A completely different situation arises when the gap value approaches the Fermi energy. This is the strong coupling case we refer to here. It is not specific to any particular pairing mechanism — it could be through a very strong electron-phonon interaction, but through any other mechanism just as well. What is common to all of them, however, is that they are attractive pairing mechanisms. In that sense the strong coupling case is still an extension of the BCS theory. The simple picture one then has in mind is that the strong coupling limit pairs form at high temperatures and condense into the superfluid state at the critical temperature. This is a Bose–Einstein condensation. The expression "BCS to BE crossover" was given by Leggett and by Nozieres and Schmitt-Rink, who have shown that the condensate is of a similar nature in both cases, and that the transition at the critical temperature evolves smoothly between a BCS condensation where pairs form and condense at the same temperature, and a BE condensation where these two steps are distinct.

A close examination of the BCS to BE cross-over has become the focus of much interest recently because in some of the cuprates measured gap values reach as we have seen several 10 meV, while the Fermi energy is a few 100 meV. We are thus very far from the case of low temperature superconductors where the two scales differ typically by four orders of magnitude. In the cuprates, the two scales do not differ by more than one order of magnitude. Is the condensation still of the BCS type, or has it already crossed over to BE?

Gap and critical temperature in the cross-over region

The relation between the gap and the critical temperature is one of the corner stones of the BCS theory:

$$2\Delta = 3.5 k_B T_c \tag{8.10}$$

Assume that we can increase the strength of the attractive pairing mechanism at will. In the strong coupling limit $2\Delta > E_F$, the Fermi sea is unstable at temperatures $k_B T < 2\Delta$ against the formation of pairs. These pairs will condense in the superfluid state at some temperature determined by the BE mode of condensation, for which the pair (boson) density is the important parameter rather than the gap. The connection between the gap and the critical temperature is then lost. Everything else being equal (the Fermi energy, the carrier density and their effective mass), as we turn on the strength of the attractive interaction the variation of the critical temperature with the gap is at first linear but eventually saturates at the BE value.

This is of course a very schematic view of what might really happen when the strength of interaction is turned on. We have no reason to assume that the electronic band structure will remain the same (the Fermi energy, the carrier effective mass and so on). Yet the main point that we have made here — the loss of proportionality between the gap and the critical temperature — should hold true. In terms of the critical temperature, there is nothing to be gained by increasing more and more the strength of the interaction. The maximum critical temperature is that which is allowed by the carrier density, their effective mass and the effective dimensionality. A good rule of thumb is that the maximum critical temperature is about a tenth of the Fermi energy. For a realistic value of the Fermi energy of a few 100 meV, this gives us a maximum T_c of a few 100K.

Change of the electrons kinetic energy upon condensation

In a normal metal, electronic states above the Fermi wave vector k_F can be occupied only by thermal excitations. This is not the case in the BCS condensate. The probability for finding electrons occupying states at $k > k_F$ decreases over a range $(k - k_F) \lesssim (\Delta/\hbar k_F)$. Hence, upon condensation from the normal state into the superfluid state, there is an *increase* of the kinetic energy, of the order of (Δ^2/E_F). The relative change compared to the Fermi energy (or the average electronic kinetic energy) is on the order of $(\Delta/E_F)^2$. In a low temperature superconductor ($\Delta \approx 1$meV,$E_F \approx 10$meV) , this change is of the order of $1{\cdot}10^{-6}$, much too small to be measurable, and indeed it was never measured.

But in the cuprates ($\Delta \approx$ a few 10 meV, $E_F \approx$ a few 100 eV), it is of the order of 1%. This change is accessible experimentally. In the compound $Bi_2Sr_2CaCu_2O_{8+\delta}$ it has been recently determined through a measurement of the real part of the conductivity $\sigma_1(\omega)$. In a single band model, the spectral weight is proportional to the kinetic energy:

$$\int_0^\infty \sigma_1(\omega) d\omega \propto -E_{\text{kin}} \tag{8.11}$$

Choice of the limits of integration is set at low frequencies by the availability of data and at high frequencies in order to eliminate interband transitions. The procedure used to obtain the relative change of the kinetic energy upon condensation consists in measuring the spectral weight in the normal state as a function of temperature, to extrapolate the temperature dependence down to zero temperature, and to compare the extrapolated value to that measured in the superconducting state, after correcting it for the spectral weight of the condensate. Changes in the value of the integral can be measured with an accuracy of a fraction of 1%.

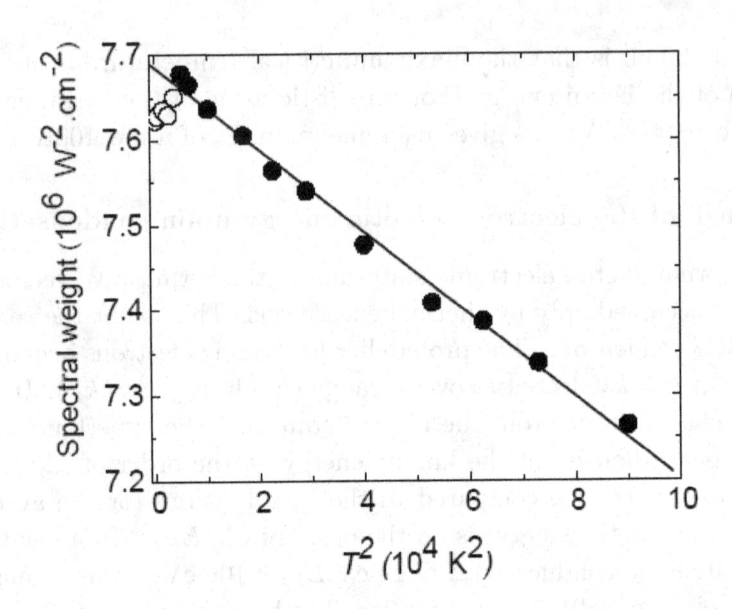

Figure 8.2: The spectral weight (Eq. (8.11)) in an overdoped Bi 2212 sample in the normal state varies linearly as a function of the temperature squared. It decrease below T_c corresponds to an increase of the kinetic energy. After G. Deutscher, A.F. Santander and N. Bontemps, Phys. Rev. B **72**, 092504 (2005).

Results of this procedure for an overdoped sample are shown in Fig. 8.2. The extrapolation is made easy by the fact that in the normal state the spectral weight varies linearly with T^2, increasing as the temperature goes down, which corresponds to a decrease in the kinetic energy. The spectral weight in the condensed state is lower than that in the extrapolated normal state value (by about 1%), which corresponds to an increase in the kinetic energy, as predicted by the BCS theory. The BCS expression for the difference in kinetic energies is (for derivation of this result see de Gennes in further reading):

$$\Delta E_{\text{kin}} = \frac{\Delta^2}{V} - \frac{N(0)\Delta^2}{2} \tag{8.12}$$

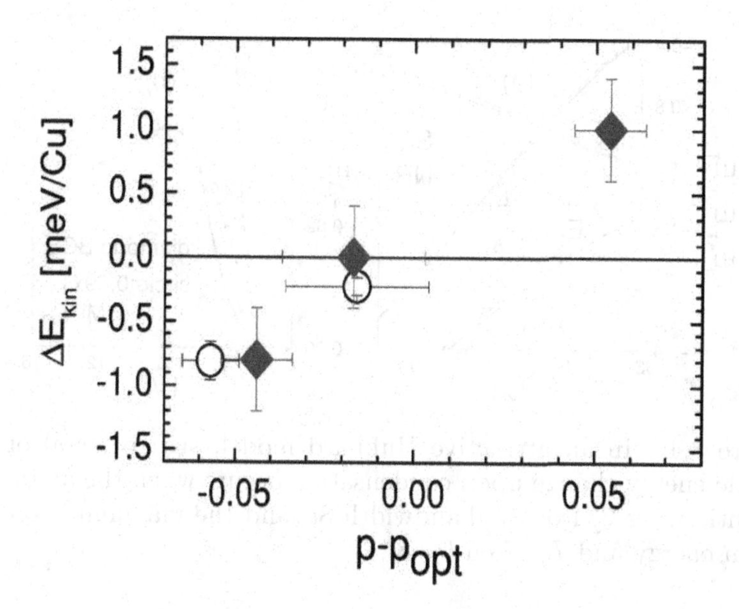

Figure 8.3: Change of the electrons kinetic energy upon condensation in the superfluid state. Note the change in sign at about optimum doping. After G. Deutscher, A.F. Santander and N. Bontemps, Phys. Rev. B **72**, 092504 (2005).

which, neglecting the second term since $N(0)V \ll 2$, gives for the relative change:

$$\frac{\Delta E_{\mathrm{kin}}}{E_{\mathrm{kin}}} \simeq \frac{1}{N(0)V}\left(\frac{\Delta}{E_F}\right)^2 \tag{8.13}$$

The measured change corresponds to $N(0)V \simeq 0.2$. It is therefore compatible with BCS, both in sign and in size.

For underdoped samples, the change is of the opposite sign, indicating that the kinetic energy diminishes in the condensed state. This is incompatible with BCS, but compatible with a BE condensation. The sign reversal occurs around optimal doping, as seen in Fig. 8.3.

Kinetic and potential energies in the normal and superconducting states have been calculated as a function of the coupling strength by

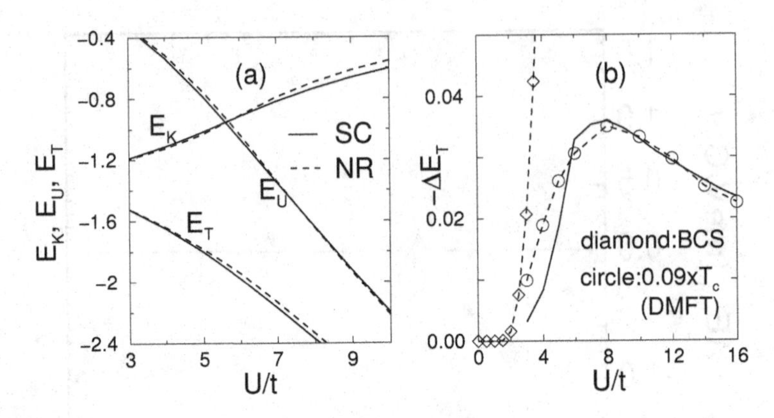

Figure 8.4: In an attractive Hubbard model, sign reversal of the kinetic energy change upon condensation occurs when the attractive potential U is 3/4 of the bandwidth $8t$, and the maximum condensation energy and T_c when U=$8t$.

Ryung *et al.*, assuming an on-site attraction leading to pair formation in the normal state for strong coupling (attractive Hubbard model), see Fig. 8.4.

In that model, sign reversal of the kinetic energy change occurs around optimum coupling (the coupling giving the highest critical temperature). Decrease of the critical temperature at very strong coupling is due to a decreasing condensation energy, itself a result of the diminishing difference between the density of states in the normal and condensed states. Both of them are characterized by a dip in the energy range corresponding to pair breaking, leading to a pseudo-gap in the normal state (with states within the gap), and a gap in the condensed state (Fig. 8.5).

Since, as shown by Leggett, the nature of the BCS and BE condensates is the same, the unconventional behavior of underdoped samples must be ascribed to an anomalous normal state. This is expected if there exist pre-formed pairs as in the attractive Hubbard model. But this is not the only possibility. As discussed in the next section, unconventional pairing mechanisms also lead to an anomalous normal state characterized by a pseudo-gap (Fig. 8.6). What

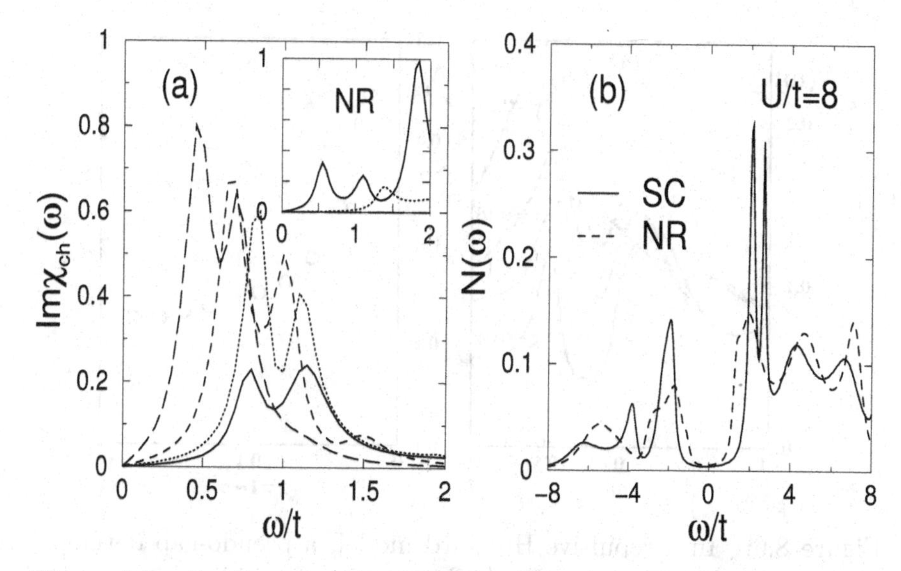

Figure 8.5: In the attractive Hubbard model at $U{=}8t$, a pseudo-gap of $2t$ develops above T_c, and transforms into a superconducting gap with coherence peaks below T_c.

is common to both conventional and unconventional schemes is that the anomalous normal state is a strong coupling effect.

Therefore, the sign inversion seen in the change in the kinetic energy upon condensation must in any case be interpreted as a strong coupling effect, in the sense that there exists an interaction potential of the order of the Fermi energy.

8.2.5 Unconventional pairing mechanisms

The models briefly reviewed above (strong electron-phonon interaction, peaks in the DOS resulting from a low dimensionality, eventually combined with an AF state prior to doping), are all based on an attractive electron-electron interaction, presumably mediated by phonons. The BCS to BE cross-over is an extreme limit of attractive potential models. Unconventional mechanisms discard the role of phonons as being essential for superconductivity, and rather

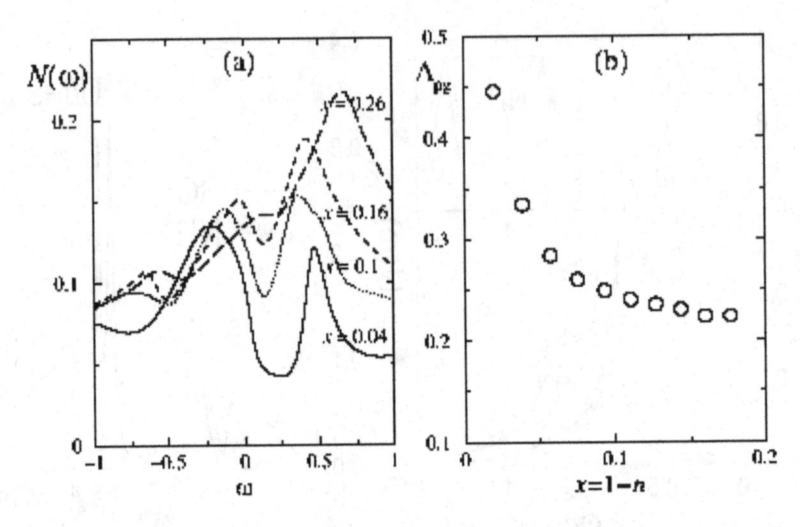

Figure 8.6: In a repulsive Hubbard model, a pseudo-gap develops in the normal state. Note the similarity with Fig. 8.1.

consider the direct repulsive electron-electron interaction as being at the origin of superconductivity. While a mechanism based on an attractive interaction can easily be pictured, a mechanism based on a repulsive one is at first view counter-intuitive. However, if it can be effective, it is immediately clear that it can lead to high critical temperatures, since the Debye energy scale is replaced by an electronic one.

The key property of the order parameter ψ function that permits a repulsive interaction mechanism of superconductivity, is that it has a phase. This phase plays no obvious role in the case of an attractive interaction, but is essential in the repulsive case. Consider the BCS self-consistent equation for the gap:

$$\Delta_k = -\sum_l V_{kl} \frac{\Delta_l}{2\left(\xi_l^2 + \Delta_l^2\right)^{1/2}} \tag{8.14}$$

where ξ_l denotes the energy in the normal state for wave vector l measured from the Fermi level, and the sum is carried out around the Fermi surface. In the BCS theory, the interaction potential is taken as negative (attractive), and the gap Δ has the same sign all around the Fermi surface. But if V is repulsive (positive), a non-trivial solution of the gap equation can still be found on the condition that Δ *will change sign around the Fermi surface.*

One such solution has the $d_{x^2-y^2}$ symmetry which has a periodicity of π:

$$\Delta(\varphi) = \Delta \cos 2\varphi \tag{8.15}$$

This solution is particularly interesting if the absolute value of V_{kl} is large for states kl such that the phase has changed sign.

The above argument does not prove that a repulsive interaction does lead to superconductivity, but it proves that if it does the phase of the order parameter must necessarily change around the Fermi surface so as to give a change in sign of Δ. As we have seen, such a change in sign has been observed in the cuprates. But it has also been pointed out (Friedel and Kohmoto) that a change in sign can occur even if the potential is attractive, due to anti-ferromagnetic fluctuations. A more decisive test of the sign of the interaction is whether the symmetry of the order parameter does or does not change across the phase diagram. Models based on a repulsive interaction give a d-wave symmetry at all doping levels. For a recent overview of the one band Hubbard model of High T_c, see Anderson in further reading. On the other hand, models based on an attractive interaction can give a mixed symmetry ($d_{x^2-y^2} + is$ or $d_{x^2-y^2} + id_{xy}$) in parts of the diagram (overdoped region). Bussmann-Holder, Keller and Muller have developed a model that combines the proximity of a Mott-Hubbard insulating state and strong lattice (Jahn Teller) distortions (see Bussmann Holder, Keller and Muller in further reading). Experiments aiming at checking the order parameter symmetry at various doping levels have so far given mixed results, as we have seen in Chapter 5. While the study of ASJ bound states show a mixed symmetry in overdoped samples, the tri-crystal experiments of Tsuei and Kirtley do not show any change as a function of doping. A final conclusion has not yet been reached on this important question.

Various models of repulsive interaction combine differently charge and spin effects. One common feature of these models is the existence of a pseudo-gap in the normal state. In that case, it is not due to the formation of pairs in the normal state, but presumably to some sort of insulating state, with electrons being kept apart and localized on lattice sites. This state may turn at low temperatures into a superfluid state having a d-wave symmetry. The evidence for this scenario comes mostly from numerical solutions such as that of the repulsive Hubbard model (Fig. 8.6). In a sense, in these models the high temperature pseudo-gap is a precursor of the superconducting state, but it is not a manifestation of the existence of pre-formed pairs. Numerical results are available for the normal state density of states in the repulsive Hubbard model, but not in the condensed state, which makes a direct comparison with experiments impossible at the moment. Likewise, no results are available for the kinetic and potential energy across the phase diagram (here the parameter is the doping rather than the strength of the potential). Therefore, one cannot say whether repulsive models are compatible with the kinetic energy measurements described above. It is *a priori* not obvious why in these models a sign reversal should occur at optimum doping.

8.2.6 Is there a BCS to BE cross-over in the cuprates?

Since the ratio $\left(\frac{\Delta}{E_F}\right)$ for YBCO and the bismuthates is of the order of 0.1, these compounds must be close to the strong coupling case. Some of the changes observed across the phase diagram are compatible with a weak coupling to strong coupling cross-over in YBCO: the absence of a pseudo-gap in overdoped samples, and its presence in underdoped ones; the fit of the heat capacity transition to a 3D XY BCS case in optimally doped samples, and not in the underdoped ones; the saturation of the critical temperature in the overdoped regime at high superfluid densities in YBCO. The sign reversal of the kinetic energy change upon condensation in Bi-2212 may be another indication for a cross-over.

The difference between YBCO and the more two-dimensional bismuthates is interesting. At optimum doping there is still a pseudo-gap in Bi 2212. T_c keeps increasing beyond that of YBCO as the superfluid density is increased, which means that the BCS regime has not been reached (see Fig. 2.6). This is in agreement with the fact that at optimum doping, the gap in Bi 2212 (30 meV) is higher than in YBCO (20meV): a higher superfluid density would be required in the bismuthates to reach their maximum critical temperature. There are thus many indications that the interaction strength in the bismuthates is higher than in YBCO. If the value of the penetration depth in the bismuthates could be reduced down to values reached in overdoped YBCO (1300 Å), the BCS regime might be reached and a critical temperature of 150K achieved at saturation. The practical importance of reaching the BCS regime is underlined in Sec. 8.4.

8.2.7 Inhomogeneous superconductivity

All theoretical models briefly reviewed above treat the superconductor as being homogeneous. There is strong experimental evidence that HTS are in fact inhomogeneous, particularly in the underdoped regime. EPR measurements have shown at low temperature a split of the line which has been interpreted as revealing the existence of hole-rich and hole-poor regions (A. Shengelaya *et al.*, Phys. Rev. Lett. **93**, 017001 (2004)). Hole-rich regions persist at very low average hole doping, suggesting the self-organization of localized holes into clusters. Second derivative STM tunneling measurements have revealed that in $Bi_2Sr_2CaCu_2O_{8+\delta}$, on the atomic scale, the electron-boson interaction is strongly inhomogeneous (Jinho Lee *et al.*, to be published). The size of hole-rich regions appears to be on the nanoscale.

A description of the cuprates as composed of small metallic clusters reminds one, of course, of the structure of granular Aluminum. In both cases, individual clusters are too small to be superconducting by themselves, since the splitting of the electronic levels is larger than the gap. Hence the superconducting state cannot be viewed as resulting from a Josephson coupling between regions in space where superconductivity is well established. We have also underlined in Chapter 4 the similarities between normal state transport properties in granular and HTS superconductors, particularly the way in which they vary across the phase diagram with the weak insulator behavior seen in both cases around and below the composition corresponding to the maximum critical temperature.

While the way in which phase separation is achieved is quite different in both cases (eutectic phase separation in granular Aluminum, possibly polaronic clusters in the cuprates, as has been suggested by A. K. Mueller and D. Mihailovic), it could be that in the end it is the small elementary cluster size that is responsible for the enhanced or high T_c. This is, of course, only a conjecture. Much remains to be done to understand normal state transport and superconductivity in these systems.

8.2.8 Critical currents in weak fields: the depairing limit

For some applications, such as cables, the important property of the superconductor is its critical current density in the weak self-field generated by the current in the conductor. Currents carried by a superconducting cable are of the order 1000Å, producing a self-field of the order of a few 0.01T. Although small this field is large enough to create some vortices, whose motion induces dissipation. But leaving this effect aside, we calculate the critical current density using the relation derived in Chapter 1 in the low temperature limit:

$$j_c(0) = ne\frac{\Delta(0)}{p_F} \tag{8.16}$$

where n is the carrier density and p_F the Fermi momentum.

If we associate $\Delta(0)$ to a maximum useful critical temperature of 200K (see Sec. 8.4), or $\Delta(0)$=40 meV, and take for n a value one order of magnitude lower than that which characterizes usual metals i.e. $5{\cdot}10^{21}/\mathrm{cm}^3$, and k_F=1Å^{-1} we get:

$$j_c(0)_{\max} \approx 5 \cdot 10^8 \mathrm{A/cm}^2 \tag{8.17}$$

Using the GL approximation:

$$j_c(t) = j_c(0)\,(1-t)^{3/2} \tag{8.18}$$

where $t = T/T_c$, we have for our ideal superconductor operating at 2/3 of its critical temperature (say at 130K):

$$j_c(130\mathrm{K}) \approx 1 \cdot 10^8 \mathrm{A/cm}^2 \tag{8.19}$$

For the canonical YBCO operating at 77K, the depairing limit is $1.6 \cdot 10^7 \mathrm{A/cm}^2$. Critical currents of up to $5 \cdot 10^6 \mathrm{A/cm}^2$ have in fact been measured in that compound, for details see next chapter.

Engineering critical currents

The above estimates are many orders of magnitude higher than the current density used in Cu wires, which is of about $100\mathrm{A/cm}^2$. (Incidentally, that value is limited by the damage that a higher current density would cause to the insulator due to the rise in temperature.) But practical superconductors are actually composites with a substantial fraction being a non-superconducting metal. The structure of these composites in described in the next chapter. The superconducting fraction f_c may range from about 1% (YBCO tape) to several 10% (BSCCO wires produced by the Powder In Tube or PIT method). The composite is called the conductor. It is often given the shape of a tape. For the engineers, the relevant critical current is then the critical current per cm-width (A/cm-width) rather than the critical current density. It is given by:

$$I_c(\mathrm{A/cm} - \mathrm{width}) = j_{\mathrm{CE}}.t \tag{8.20}$$

where j_{CE} is the engineering critical current density (the critical current density of the composite conductor) and t the thickness of the tape. Taking j_c=$1 \cdot 10^8 \mathrm{A/cm}^2$ as for our optimum superconductor, f_c=0.01, t=0.01 cm, we get:

$$I_{c\,\mathrm{max}} \approx 1 \cdot 10^4 \mathrm{A/cm} - \mathrm{width} \tag{8.21}$$

8.3 Upper critical fields

8.3.1 Zero temperature limit

At temperatures $T \ll T_c$, where we can ignore fluctuation effects, superconductivity can be quenched either by the orbital effect that determines H_{c2}, or by a spin effect. This is the so-called Pauli limit H_p, where the order parameter is quenched when the splitting between the electronic levels induced by the field reaches the gap Δ

(more exactly when $\mu_0 H$=0.7Δ):

$$H_{c2} = \frac{\Phi_0}{2\pi\xi^2}; H_p = 18,000 T_c \tag{8.22}$$

where H_p has been calculated using the BCS relation between Δ and T_c. Here H_p is expressed in Tesla and T_c in degrees Kelvin.

The value of H_{c2} can be enhanced by reducing the normal state mean free path l below the clean limit coherence length ξ_0. In the dirty limit $\xi_0 \ll l, \xi \approx (\xi_0 l)^{1/2}$. The lowest value of the mean free path compatible with a metallic state is given by the Joffe-Regel rule $(k_F l) \geq 1$. In practice, when scattering is introduced by impurities in solid solution in the metal, it is difficult to reduce l below 10Å. For instance, starting from pure Nb (ξ_0=400Å) and introducing Ti impurities, we can get a coherence length of about 60Å, and an orbital critical field of 10T. For Nb, the Pauli limit is 17T. For Nb_3Sn (T_c=17K), the Pauli limit is 30T, and H_{c2} reaches above 20T. Spin orbit scattering removes in part the Pauli limit, but this limit is still useful as the order of magnitude of the achievable upper critical fields.

The above examples show that upper critical fields in excess of 100T can only be reached with High T_c materials. With T_c=100K, H_p=180T. An upper critical field in excess of 100T has been measured in YBCO at low temperature, less than a factor of 2 smaller than H_p. For our hypothetical optimized superconductor having T_c=200K, H_p=360T.

Problems related to the short coherence length

The short coherence length of the HTS, which allows reaching extremely high fields, has also its disadvantages. Besides reducing the coherence volume as discussed above, it also makes the order parameter very sensitive to crystallographic defects. Even minor ones, such as twin boundaries or local changes in the doping level on the atomic scale, can locally lower it. This can be easily understood. In a low T_c — long coherence length — material, the influence of an atomic size defect is averaged out on the relevant scale of ξ and it

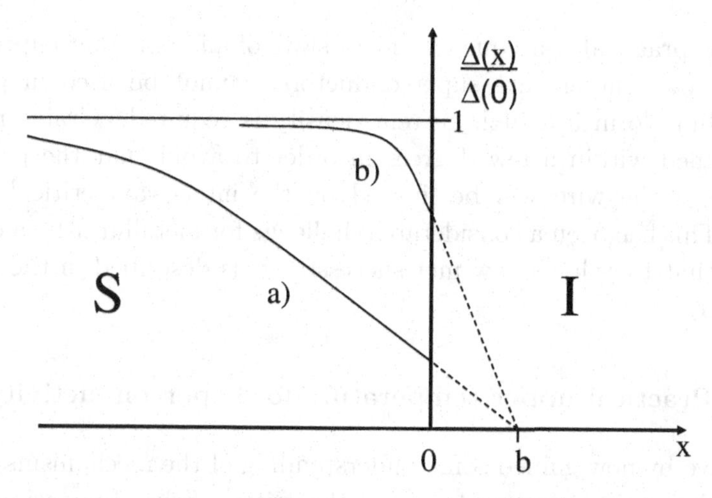

Figure 8.7: In a short coherence length superconductor, the order parameter near a free surface bends down substantially. Curves a and b are respectively for high and low temperatures. After G. Deutscher and A.K. Muller, Phys. Rev. Lett. **59**, 1745 (1987).

has a negligible effect. This is not the case if ξ is about two or three lattice spacings, as it is in the cuprates. For instance, at the surface the pair potential goes to zero on the atomic scale a. When $\xi \gg a$, *on the scale of* ξ the order parameter is unaffected by the presence of the surface. Evidently, this is not the case if $\xi \approx 3a$ as happens in the HTS (Fig. 8.7). More generally, one can define an extrapolation length scale $b = \left[\Delta\left(\frac{d\Delta}{dx}\right)\right]^{-1}$ where the gradient is taken in the direction perpendicular to the surface, and show that it is of the of order (ξ^2/a). When $\xi \gg a$, $b \gg \xi$, and the usual zero gradient approximation is valid.

The short coherence length is the fundamental reason why in the HTS the order parameter is depressed at grain boundaries, and the critical current density across a grain boundary much smaller than what it is in the bulk. The details of what exactly goes on at and near the boundary are still imperfectly understood (see Hilgenkamp and Mannhart for further reading), but the weak link character of grain boundaries is well established and universal in the cuprates.

The practical consequence of this state of affairs is that cuprates, unlike low temperature superconductors, cannot be used in polycrystalline form if a high current density is required. Grains must be aligned within a few degrees in order to avoid that the performance of the wire will be limited by the inter-grain critical current. This has been a considerable challenge for metallurgists, a challenge that they have now met successfully, as described in the next chapter.

8.4 Practical upper temperature for superconductivity

We have by now gained some understanding of the mechanisms that limit the critical temperature and the critical field. Increasing the interaction strength beyond a certain value does not help, because the superfluid density then becomes the limiting factor. The short coherence length necessary to obtain high critical fields introduces severe constraints on the material perfection.

A somewhat distinct point is that of the practical properties of the superconductor at high temperatures. Today, superconducting tapes are manufactured from two different cuprates, Bi 2223 which has a maximum critical temperature of 125K and YBCO of 92K (for more details on these tapes, see next chapter). One might naively have guessed that at liquid Nitrogen temperature, the performance of the Bi tape should be far superior to that of YBCO, since at that temperature it is further away from its transition. But in fact, this is not the case. At that temperature the critical current of Bi tape is strongly reduced by weak magnetic fields (say 0.1T), and is far inferior to that of YBCO tape. Under magnetic fields of more than 1 Tesla, Bi tape can only be used below 30K. The empirical evidence is thus that for practical use the critical temperature is not the only parameter that determines the material performance at $T \lesssim T_c$.

8.4.1 Role of the condensation energy

We have underlined in preceding chapters the important role of the condensation energy per coherence volume U. Basically, it must be larger than k_BT for vortex pinning to be effective. The key parameter is the ratio $(U(0)/k_BT_c)$. It must be equal to or larger than one to allow an effective vortex pinning at temperatures of the order of k_BT_c.

As shown in Chapter 1, in the weak coupling limit the condensation energy per unit volume is given (in the clean limit and in the isotropic case) by:

$$U(0) = \frac{2}{\pi^5}\frac{E_F^2}{\Delta} \tag{8.23}$$

based on the coherence volume ξ^3.

For the quasi-2D anisotropic case the coherence volume is given by:

$$\Omega_c = \xi_{ab}^2 \xi_c \tag{8.24}$$

and we get instead:

$$U(0) = \frac{2}{\pi^5}\frac{E_F^2}{\Delta} r \tag{8.25}$$

where $r = (\xi_c/\xi_{ab})$.

Another, equivalent one is:

$$U(0) = \frac{E_F}{\pi^4}(k_F \xi_{ab}) \tag{8.26}$$

Let us assume that we can, by a clever trick, increase the critical temperature of a BCS superconductor. What would be the highest useful value of T_c?

From Eq. (8.23), the condition $U(0) > k_BT_c$. reads:

$$k_BT_c < \left(\frac{2}{\pi^5}\right)\frac{(E_F)^2}{\Delta} \tag{8.27}$$

According to BCS theory, $2k_BT_c \sim \Delta$, hence this condition is equivalent to:

$$\Delta < \left(\frac{4}{\pi^5}\right)^{\frac{1}{2}} E_F \tag{8.28}$$

For a broad band superconductor, $E_F \geq 1\text{eV}$, the value of Δ should be limited to about 100 meV, or $T_c \lesssim 500\text{K}$. Such a high temperature superconductor, certainly the best we could hope to get for practical applications, could be used at room temperature.

Any realistic scheme to reach a high critical temperature requires a reduced dimensionality. Including the anisotropic ratio, the "usefulness" condition reads:

$$\Delta < r \left(\frac{4}{\pi^5} \right)^{\frac{1}{2}} E_F \tag{8.29}$$

Given the Fermi energy, this relation introduces a link between the anisotropy (basically a normal state property) and the maximum useful gap value. Should it exceed that value, the practical properties of the superconductor would in fact deteriorate. Let us see how this prediction fits the known properties of the cuprates. For YBCO, the least anisotropic of the cuprates, r=0.2. For the more typical cuprate BSCCO, r=0.03. These values are for optimally doped samples. The quasi 2D character reduces the maximum useful value of Δ from 100 meV to 20 meV for YBCO and to a few meV for BSCCO. For YBCO, the maximum useful value of the gap matches the measured one. One can expect useful pinning properties up to temperatures of the order of T_c. But for BSCCO, the measured gap (30 meV) is much larger than the maximum useful one (3 meV). In other words, it is only below about 30K that one can expect pinning to become effective. This is in line with experimental results. In spite of its higher critical temperature, BSCCO is not as good a practical superconductor as YBCO. Its energy gap is too large compared to its Fermi energy, when we take into account the anisotropy (Eq. (8.29)); or, in other terms, its condensation energy per unit volume is too small compared to $k_B T_c$.

A different approach to estimate the value of $U(0)$ is to use the Ginzburg–Landau theory. The advantage of this approach is that it is somewhat more general than the BCS one. In particular, we do not need to assume the BCS expression for the condensation per unit volume to be valid as, indeed, it does not hold for a number of cuprates. GL theory holds for dirty as well as for clean superconductors. It is valid in the presence of magnetic fields and generally in all situations where the order parameter is small. This is useful when we want to estimate in what range of temperatures the superconductor develops useful properties.

We start from the expression for the condensation energy in terms of the thermodynamical critical field H_c:

$$\Delta F = \frac{(H_c)^2}{8\pi} \tag{8.30}$$

which can be rewritten as:

$$\Delta F = \frac{1}{8\pi}\left(\frac{\Phi_0}{2\pi\sqrt{2}\lambda\xi}\right)^2 \tag{8.31}$$

Taking into account the anisotropy factor, the condensation energy per coherence length reads:

$$U(0) = \frac{r}{64\pi^3}\frac{(\Phi_0)^2}{\lambda_{ab}^2}\xi_{ab} \tag{8.32}$$

The advantage of using this expression to calculate U is that we only need to plug in experimental values for λ, ξ and r. These parameters are more readily available than values of the condensation energy, the Fermi energy and the gap.

For YBCO at optimum doping, the relevant parameters are λ =1300Å, ξ=15Å,r=0.2. We obtain for $U(0)$ a value of the order of 100K, close to the value of T_c. This confirms the result that we have obtained using the BCS approach: it applies marginally to this material.

For the more typical cuprate BSCCO, using the same values for λ and ξ (in fact, λ has been measured to be somewhat larger in BSCCO than in YBCO, and ξ somewhat shorter), but taking into account the smaller anisotropy ratio r=0.03, we obtain for $U(0)$ a value much smaller than T_c, about 20K. This confirms that this cuprate can be useful in strong fields only at temperatures well below T_c, and that the BCS theory is basically inappropriate to describe its properties.

These examples have served to illustrate the general idea that for a given carrier concentration or superfluid density, and a given anisotropy ratio there exists an optimum value of the gap beyond which the practical superconducting properties actually deteriorate.

While the condensation energy per unit volume is a directly measurable quantity, for instance from heat capacity measurements, the coherence volume is not. It is therefore important to understand how it varies when going from the weak coupling to the strong coupling regime.

The variation of the coherence length as a function of the strong coupling parameter (Δ/E_F) has been calculated by Pistolesi and Strinati. As expected, in the weak coupling limit the coherence length is inversely proportional to the gap. In the strong coupling limit it becomes necessary to distinguish between the phase coherence length (which describes for instance the spatial recovery of the order parameter when moving away from a vortex center), and the pair size. As the coupling strength is increased, these two length scales depart from one another, with the pair size saturating at a low value and the phase coherence length going back up in the strong coupling regime (Fig. 8.8). The coupling strength at which ξ_{phase} reaches a minimum signals the weak coupling to strong coupling cross-over. This is an unfavorable region for applications. The bismuthates may be close to it.

8.4.2 Loss of line tension

In Chapter 7, we have reviewed the experimental evidence for melting of the flux line lattice in the HTS at a field $H_m < H_{c2}$, and noted that flux pinning beyond H_m requires pinning of the individual vortices. This is possible as long as vortices retain their line tension, otherwise they will easily deform and escape from pinning defects. In this section, we wish to investigate up to what maximum field H_L their rigidity can be maintained. The field H_L is the maximum value that the irreversibility field H_{irr} can reach, if the appropriate pinning defects can be engineered. The value of $H_L(T)$ is of considerable importance because it determines the maximum temperature up to which one can hope to operate a device such as a motor or generator

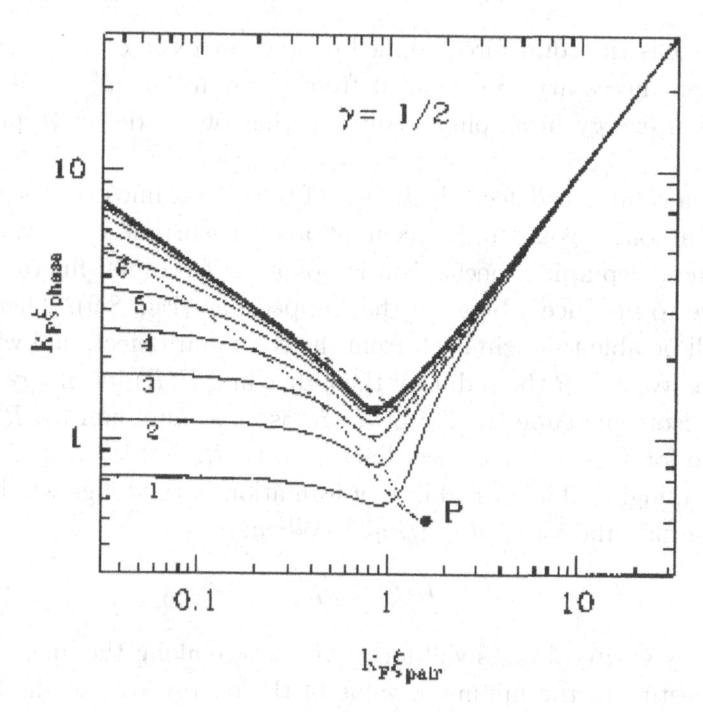

Figure 8.8: The pair size ξ_{pair} and the phase coherence length ξ_{phase} become distinct in the strong coupling limit $k_F\xi_{\text{pair}}$¡1. Curves are for different carrier densities, the thicker one being the low density limit. After F. Pistolesi and G. Strinati, Phys. Rev. B **53**, 15168 (1996).

at the field value (typically a few Tesla) required for engineering and economical reasons.

The most favorable pinning defect is a columnar one, since the vortex line can be held in place over its entire length as long as it remains a rigid entity. The field H_L is therefore the relevant one for this "ultimate defect".

For a pinning center, depinning occurs when the Lorentz force applied on the vortex over a certain length is sufficient to extract the line from the potential well where it is trapped:

$$F_L \cdot w = E_c \tag{8.33}$$

where F_L is the total force applied on the vortex line, w is the displacement necessary to extract it from the well, and E_c is the condensation energy in a volume equal to that of the defect (typically ξ^3).

For a columnar defect, depinning of the vortex line over its entire length at once would only occur at a huge current density. But a different depinning mechanism is possible, if thermal fluctuations are able to produce a bulge in the trapped line (Fig. 8.9). Then the line will be able to unzip itself from the columnar defect, and will be set free. We call R the radius of the bulge, and $U(R)$ its energy cost. For an isotropic superconductor, $U(R)$ is at a minimum for $R \approx \xi$, because for $R \gg \xi$ it increases linearly with R, and for $R \ll \xi$, it is not meaningful. The probability of formation of the bulge will be an exponential function of $[U(\xi)/k_B T]$. When:

$$U(\xi) \approx k_B T \tag{8.34}$$

thermally excited bulges will form everywhere along the line.

To estimate the minimum value of the energy cost of the bulge we use:

$$U(\xi) \simeq \pi \xi L \tag{8.35}$$

with:

$$L = \frac{\Phi_0^2}{4\pi^2} \lambda(T,H)^{-2} \tag{8.36}$$

For a high κ value, we can use the approximation:

$$\lambda(T,H)^{-2} \simeq \lambda(T,H=0)^{-2} \frac{H_{c2} - H}{H_{c2}} \tag{8.37}$$

For an anisotropic superconductor, characterized by in-plane and out-of-plane coherence lengths ξ_{ab} and ξ_c, with the magnetic field oriented along the c-axis, the corresponding bulge having the lowest energy has extensions ξ_{ab} and ξ_c. Its energy is reduced by the factor $r = (\xi_c/\xi_{ab})$:

$$U(\xi) = rU(\xi_{ab}) \tag{8.38}$$

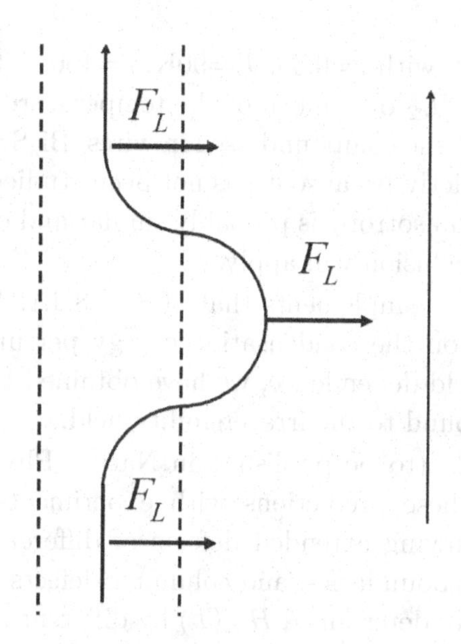

Figure 8.9: A vortex line trapped in a columnar defect can be unzipped from it under a Lorentz force if a thermally excited bulge has been formed.

Assuming the GL temperature dependencies of ξ and λ, we get from the above expressions:

$$H_L(T) \simeq H_{c2}(t)\left[1 - \alpha g t\,(1-t)^{-1/2}\right] \tag{8.39}$$

where α is a number of order unity, t is the reduced temperature and:

$$g = 40\Phi_0^{-2} r^{-1} k_B T_c \kappa \lambda(0) \tag{8.40}$$

where λ is to be taken as λ_{ab} for the anisotropic case as described above.

For the NbTi alloy used to make superconducting magnets, r=1, T_c=9.2K, κ=30, $\lambda(0)$=1,500Å, we calculate g=0.0013. For all practical purposes, $H_L \equiv H_{c2}$. Vortex lines remain rigid up to the upper critical field except in a very narrow region near T_c. For YBCO, with r=0.15, T_c=90K, κ=100, λ=1.500Å, we get g=0.08. At T=77K, H_L=0.8H_{c2}. This is still a rather favorable situation. For

$Bi_2Sr_2CaCu_2O_{8+\delta}$, with r=0.03, T_c=90K, κ=150, $\lambda(0)$=2.200 Å, we get g=0.9. $H_L \ll H_{c2}$ over much of the temperature range. The exact parameters for the compound used in wires, $Bi_2Sr_2Ca_2Cu_3O_{10+\delta}$, are not known exactly because it has not been studied in single crystal form, but the anisotropy is probably similar and one may assume that the same conclusion will apply.

This discussion complements that of Sec. 8.4.1. Starting from a similar emphasis on the condensation energy per unit volume, but adding to it its field dependence, we have obtained Eq. (8.39) which gives an upper bound to the irreversibility field.

Figueras *et al.* (to be published in Nature Physics June 2006) have compared these predictions with experiments performed on YBCO samples having extended defects of different kinds such as dislocations, twin boundaries, and columnar defects created by irradiation. They have determined $H_{\rm irr}(T)$ by different methods such as the anisotropy in the magneto-resistance, and flux transformer experiments in which the persistence of vortices across the thickness of the sample can be tested by comparing voltages induced by flux motion on two opposite faces. Their conclusion is that for all defects studied and all methods of determination of the irreversibility line, H_L represents the upper limit that $H_{\rm irr}$ can reach. At 77K, $H_{\rm irr}$ reaches up to 10 Tesla, in good agreement with our estimate for H_L at that temperature (see Fig. 8.10).

8.4.3 Concluding remarks: coupling strength versus useful high T_c

With one interesting exception, the cuprates discovered by Bednorz and Muller are the first superconductors whose critical temperature is a significant fraction of the Fermi temperature (the exception is doped $SrTiO_3$, but this compound has a very small Fermi temperature and a critical temperature in the 1K range). This realizes the old dream of having a critical temperature that is not limited by phonon frequencies. The cuprates approach the strong coupling limit, defined here as that for which the pairing gap reaches the Fermi energy of the same electron gas without interaction. High critical currents

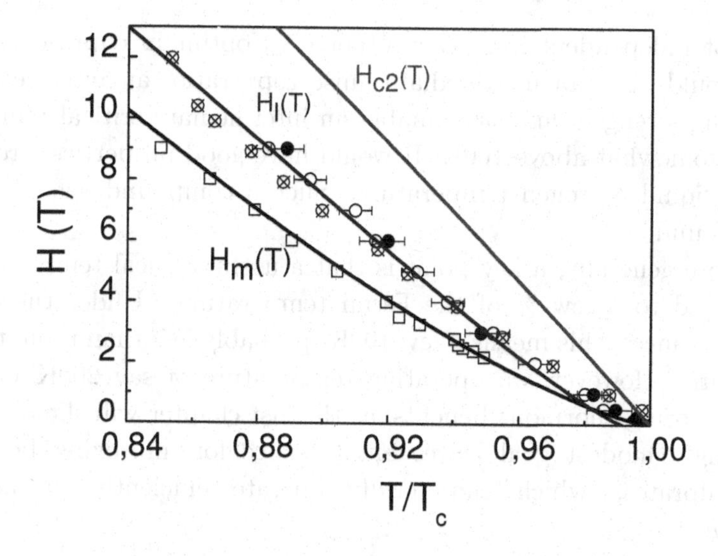

Figure 8.10: Temperature dependences of the melting field Hm, of the limiting field HL where line tension is lost and of the upper critical field H_{c2} in YBCO. The limiting field continuous line is a fit to the data using Eq. 8.39 (after J. Figueras, T. Puig, X. Obradors, W. K. Kwok, L. Paulius, G. W. Crabtree and G. Deutscher, Nature Physics June 2006).

have also been achieved, approaching in YBCO the BCS depairing limit.

We have emphasized the role played by the condensation energy per coherence volume. We have concluded that there is indeed a fundamental reason why under applied fields at liquid Nitrogen temperature the bismuthates have less useful superconducting properties than YBCO, even though they have a higher critical temperature. Their larger gap, and earlier apparition of the pseudo-gap show that they have a larger coupling strength. The condensation energy per coherence volume is reduced by the shorter ξ and the reduced density of states at the Fermi level, due to the pseudo-gap which appears to be a robust feature of strongly coupled superconductors. In addition, the bismuthates have a higher anisotropy, which further reduces the coherence volume. It is tempting to speculate that these two parameters, coupling strength and anisotropy ratio, are not independent from each other. An optimum cuprate material would have an intermediate anisotropy ratio, an intermediate coupling strength, and presumably an intermediate critical temperature somewhat above 100K. It would have good properties already above liquid Nitrogen temperature. Such a compound has not yet been found.

More generally, a key point is that a useful critical temperature is limited to a few % of the Fermi temperature. Under the best circumstances, this means a few 100K, probably less than room temperature. However, an operating temperature of say 200K would already bring enormous benefits, as the last chapter will show. This seemingly modest goal leaves ample space for improving beyond the cuprates, which can hardly operate efficiently at liquid Nitrogen.

8.5 Further reading

For the effects of a Van Hove singularity in the A15 compounds, see J. Labbe and J. Friedel, J. Phys. Radium **27**, 153, 303, (1966); for its relevance to the High T_c cuprates, see J. Labbe and J. Bok, Europhysics Lett. **3**, 1225 (1987).

For the band anti-ferromagnetic model of the cuprates, see J. Friedel and M. Kohmoto, Eur. Phys. J. B. **30**, 427 (2002).

For a review of grain boundary effects see H. Hilgenkamp and J. Mannhart, Rev. Mod. Phys. **74**, 485 (2002).

For a discussion of the effect of strong coupling on the coherence length, see F. Pistolesi and G. Strinati, Phys. Rev. B **53**, 15168 (1996).

For a recent overview of the one-band Hubbard model of High T_c, see P.A. Anderson cond-mat/0510053

For a model combining strong electron correlation effects and strong lattice distortions, see A. Bussmann-Holder, H. Keller and A.K. Muller in "Structure and Bonding" Eds. A. Bussmann-Holder and A. K. Muller, Springer-Verlag (2005).

For a derivation of the electron kinetic energy increase in a BCS condensate, see P.G. de Gennes, op. cit. Chapter 2.

Chapter 9

HTS conductors and their applications

In this chapter we present a brief overview of the manufacturing processes and applications of HTS conductors. It is intended to give the reader some appreciation of the work that has been done to transform ceramic materials into practical conductors, and to provide some perspective of what their main applications might be in the future, The emphasis is here on high performance (high j_c) wires, that can only be obtained by the elimination of large angle grain boundaries. These wires will be the basis for most HTS applications. There are also more limited but interesting applications for bulk ceramic superconductors which we now mention briefly for the sake of completeness.

The first commercial applications of HTS are in fact as current leads for LTS magnets. Below liquid Nitrogen temperature, they combine low electrical losses as superconductors and low thermal conduction losses as ceramics. They reduce considerably the rate of evaporation of the liquid Helium bath generally used to cool these magnets. All large scale LTS magnets now use such current leads. They come in the form of massive tubes of polycrystalline Bi 2212. Even a low j_c of a few 100A/cm^2 is sufficient to carry a current of the order of a few 100A in a cylinder having a cross-section of the order of 1cm^2. Similar cylinders have been used in Fault Current Limiters (Fig. 9.1). These devices act as non-destructive switches in case of a surge in current due to a fault somewhere in the network: the

Figure 9.1: A three-phase Fault Current Limiter composed of bulk 90 Bi-2212 cylindrical tubes. (Courtesy of Siemens.)

high current induces a return to the normal state, which introduces a large resistance in the circuit thus reducing the fault current down to an acceptable level.

Another application of bulk HTS is for magnetic bearings. Here the HTS is in fact in the form of YBCO quasi-single crystals having strong pinning properties. The application is a flywheel energy storage device, levitated through a combination of permanent magnets and magnetized YBCO crystals. Lateral stability of the flywheel is provided by pinning of the flux lines in the YBCO crystals. The combination of magnetic levitation and vacuum environment allows

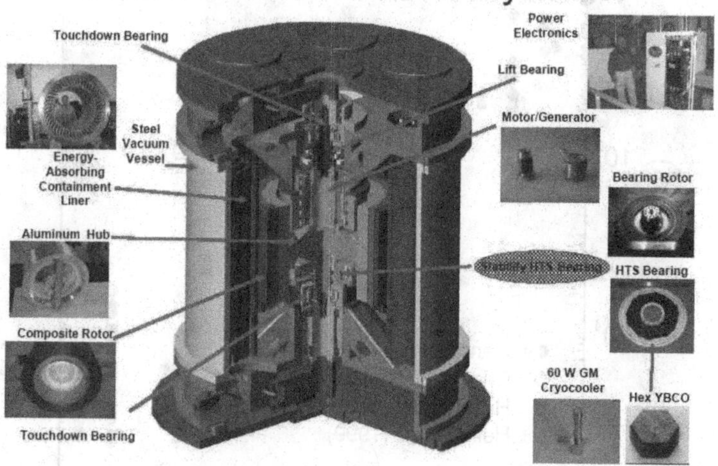

Figure 9.2: Components of a Superconducting Flywheel Energy Storage device. (Courtesy of M. Strasik, Boeing Phantom Works.)

a slow decay of the stored energy, of the order of 1% per day. Energy is pumped in and retrieved through a motor-generator. The power provided ranges from a few kW for long term storage, to several 10kW for power demand shaving, and up to a few 100kW for providing power in case of grid failure until generators come on line, or to improve power quality (Fig. 9.2).

9.1 Grain boundaries

Grain boundaries in polycrystalline cuprates are responsible for the large difference between the critical current values measured by transport and those determined from magnetization measurements: the first ones are inter-grain, and the later intra-grain values. The reason for this difference was clearly established by measurements of current-voltage characteristics across artificial grain boundaries produced by epitaxial growth of HTS films on bi-crystalline substrates. Grain Boundaries were found to behave as Josephson junctions (called "GBJ"), whose critical current falls off exponentially as a function

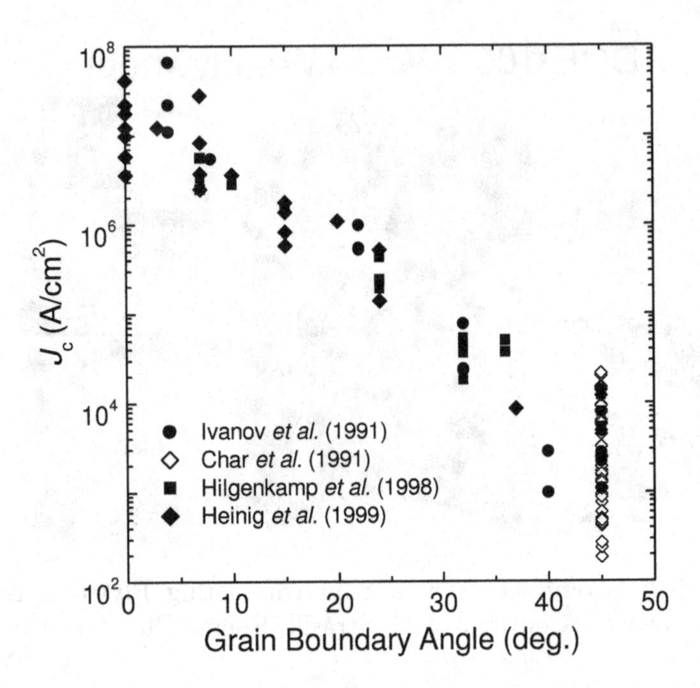

Figure 9.3: Critical current densities through tilt grain boundary junctions. After H. Hilgenkampf and J. Mannhart, Rev. Mod. Phys. **74**, 485 (2002).

of misorientation between the two sides, diminishing by two to three orders of magnitude for a 45° misorientation (Fig. 9.3).

It does not seem that the exact type of misorientation matters a lot, whether it is in-plane (the two sides having their c-axis parallel to the substrate) or out-of plane, or any combination of the two kinds. Electron microscopy studies have shown that these artificial boundaries are "clean", namely do not contain secondary insulating phases. Besides the generic sensitivity to defects due to the short coherence length, the details of why GBJ act as Josephson junctions are still incompletely understood. The I_cR_N product is also a strong function of the misorientation (Fig. 9.4).

(i) The d-wave symmetry does contribute to the decrease of j_c. The GBJ critical current is obtained by summing up over all k

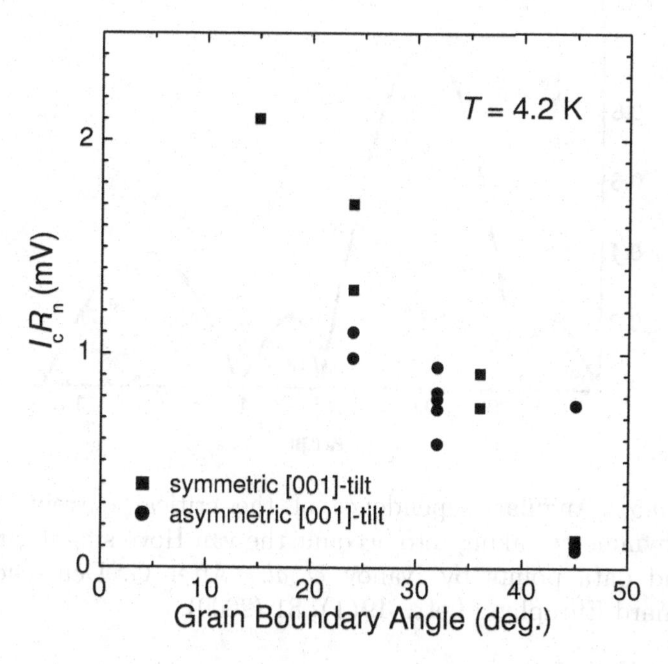

Figure 9.4: Angular dependence of the I_cR_N product of grain boundary junctions as a function of misorientation. After H. Hilgenkampf and J. Mannhart, op. cit.

vectors:

$$I_c = \sum_{k\|} \Delta_L \Delta_R$$

where Δ_L and Δ_R on the left L and on the right R are for k vectors having equal components parallel to the boundary. The decrease resulting from the misorientation is substantial, but still much weaker than the experimentally measured values (Fig. 9.5).

(ii) The G. B. is not a simple metal to metal contact. Its normal state resistance may be due in part to charging and band-bending, which reduce the carrier density (for a detailed discussion see Hilgenkampf and Mannhart, in further reading suggestions).

(iii) There are structural indications that the oxygen concentration is reduced near the boundary (for details see Y. Zhu *et al.*, Phil. Mag.

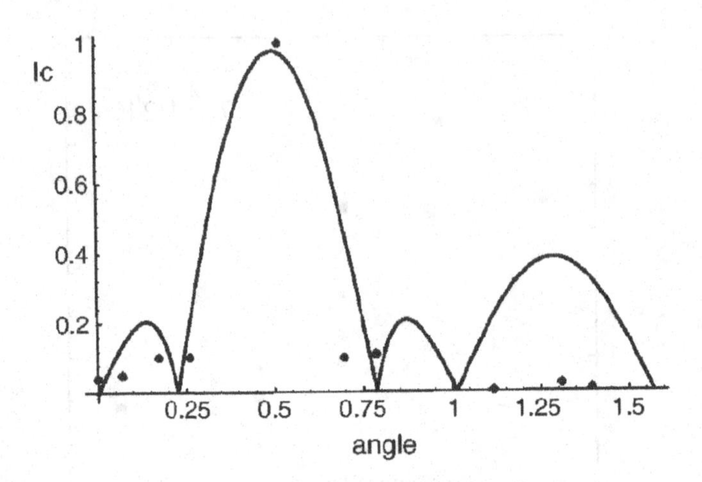

Figure 9.5: Angular dependence of the critical current due to d-wave symmetry taking into account the van Hove singularity (full line) and data points by Ivanov *et al.* After G. Deutscher and R. Maynard Europhys. Lett. **49(1)**, 81 (2000).

A **70**, 969-84 (1994)). The mechanism for hole depletion appears to be related to the strain field surrounding the array of dislocations at the boundary. Hole depletion is the way for the lattice to adjust to the local mismatch. The depleted zone may extend over distances of the order of 100Å, and was found to increase linearly with misorientation (Fig. 9.6).

A possible general model for the electronic structure of a large misorientation GB is a gradual decrease of carrier concentration when moving towards the boundary, with a corresponding decrease of the superconducting gap over a thickness d_I (Fig. 9.7). The exponential decrease of j_c with misorientation might be due to the linear increase of d_I with the misorientation (see N.D. Browning *et al.*, Physica C **294**, 183 (1998) for a discussion of this interpretation). The boundary resistance R_N and the critical current I_c of the junction are then controlled by a factor $\exp(-d_I/r_0) \propto \exp(-\Theta/\Theta_0)$ where r_0 is the decay length of the electronic wave function in I, Θ the misorientation angle and Θ_0 is determined by the properties of the barrier.

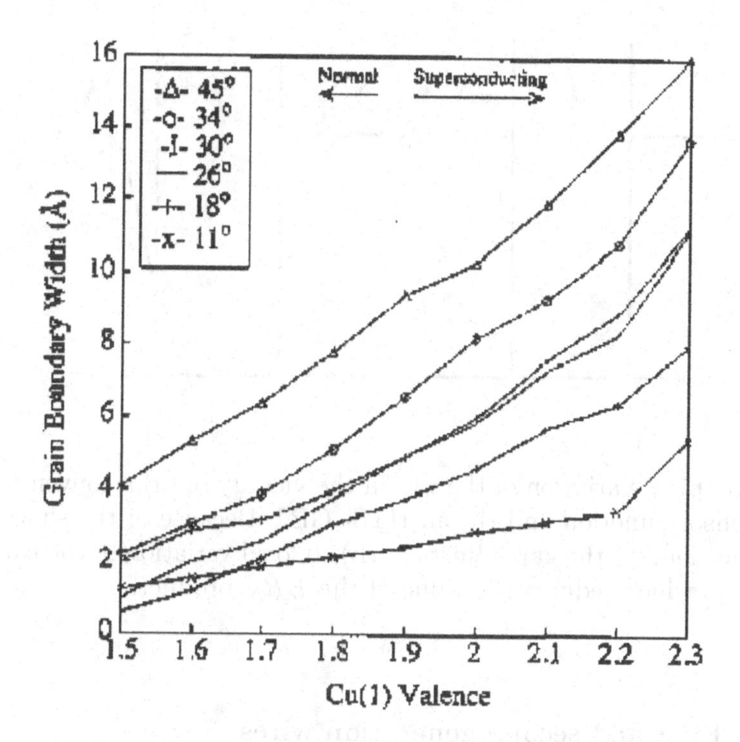

Figure 9.6: Width of the carrier depleted region in grain boundaries for different misorientations. After Browning *et al.*, op. cit.

The product $I_c R_N$ is reduced compared to the Josephson value corresponding to the full gap in the banks, because of the progressive gap reduction near the insulating region. This is the main difference between a LTS Josephson junction, where the gap remains constant in the superconducting banks up to the boundary with the dielectric, and the $I_C R_N$ product is equal to the full gap. In an HTS-GBJ, the coherence length is on the same scale as the depleted region, the gap adjusts to the local carrier concentration, and the value of $I_c R_N$ is strongly reduced. It goes from several meV for small angles down to less than 0.1 meV for a 45° angle. For small angle GBs, an S/N/S model of the junction may be more appropriate because the oxygen depletion zone is too small for an insulating region to form.

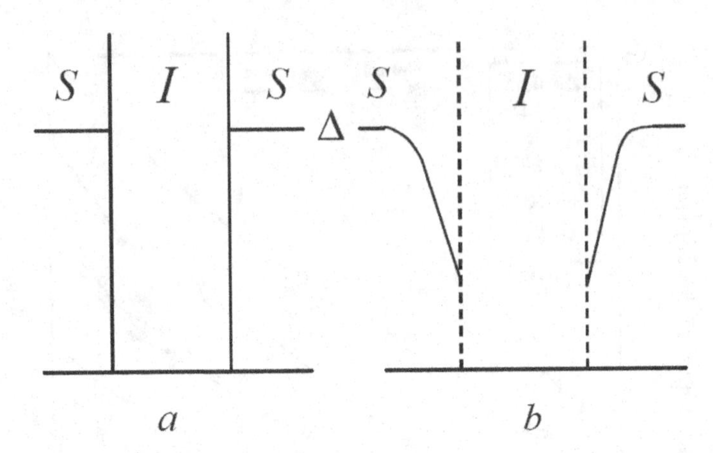

Figure 9.7: Variation of the gap in the vicinity of (a) a conventional Josephson junction and (b) an HTS- GBJ. Because of the short coherence length, the gap adjusts in (b) to local variation of the carrier density, which reduces the value of the $I_c R_N$ product.

9.2 First and second generation wires

Since making mile long single crystals is eminently impractical, methods must be found to obtain a good orientation of grains that have nucleated independently of each other. Two methods have been developed, the first one based on a nearly conventional drawing process and the other on thin film deposition.

The first method consists of introducing the HTS material in powder form in a silver tube, which is then drawn in successive steps and annealed. First generation wires are manufactured by this "Powder In Tube" or PIT method. It is restricted to Bi compounds, because they cleave easily. Cleavage takes place during drawing between neighboring BiO planes weakly bound by van der Walls forces. Platelets having their large faces parallel to CuO_2 planes are formed and slide past each other, resulting in a "brick wall" geometry. Current in the wire flows parallel to the platelets. It can pass from one platelet to another not only through the platelets edges, but also through their large faces. Multi-filamentary Bi 2223 and

Bi 2212 wires and tapes are commercially produced with individual tubes bundled together, and available in kilometer length. About two thirds of the wire volume is made of silver alloy.

The second method consists in growing HTS film on suitably oriented substrates. It is used for the YBCO or "123" family of cuprates, which does not cleave easily and cannot be produced by the "PIT" process. The reduced anisotropy of YBCO makes it as we have seen in the preceding chapter a much more desirable superconductor than BSCCO, at the same time complicates the fabrication process and makes it unavoidable to use complex thin film deposition techniques. Contact between YBCO grains is only through their edges. Reaching a high j_c then requires an excellent alignment of these grains. This alignment can only be achieved by hetero-epitaxial growth on well oriented substrates. Manufacturing of "second generation wires" or "coated conductors" may involve the successive deposition of up to half a dozen different layers. Vacuum as well as non-vacuum methods of deposition have been developed for their growth. For details see Sec. 9.2.3.

9.2.1 Properties of first generation wires

Bi 2223 wire reach engineering current densities of about 50 kA/cm^2 at 77K, which is sufficient for cable applications. But coils made of Bi 2223 wire cannot produce any field of interest at that temperature, because the critical current density collapses in fields of less than 0.1T. They can however produce fields of a few Tesla at temperatures of about 20K.

While research and development have resulted in much improved critical current of Bi 2223 wire in self-field, no significant improvement of the irreversibility field, and therefore of critical current under applied fields, has been achieved. This empirical result is in line with the theoretical expectations proposed in the preceding chapter. The fundamental reason for the poor performance of Bi 2223 under applied fields is that the condensation energy per coherence volume is very small, due in large part to the strong anisotropy. It becomes larger than $k_B T$ only at temperatures well below T_c, which precludes

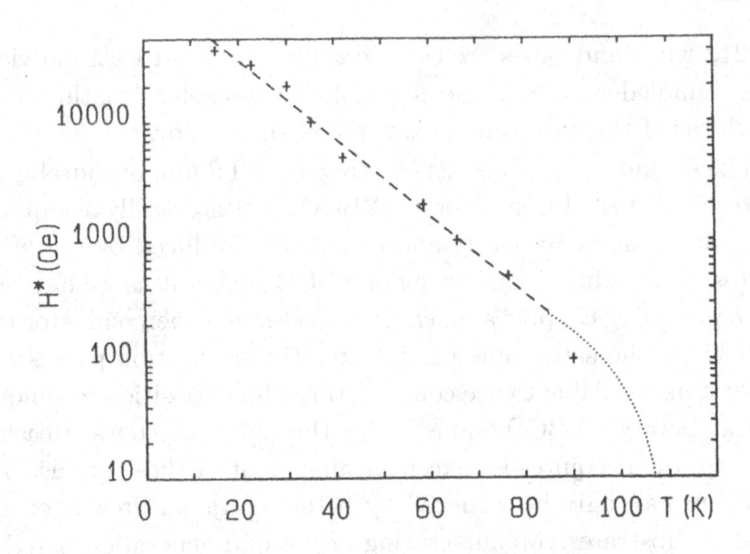

Figure 9.8: Exponential increase at low temperatures of the irreversibility field in Bi-2223. Values of the order of 1 Tesla, necessary for applications in rotating machines, are only reached below 20K. After P. de Rango *et al.* J. de Physique **50**, 2895 (1989).

efficient vortex pinning *by any kind of defect* at temperatures of the order of T_c.

The irreversibility field of Bi 2223 has been studied at low temperatures (Fig 9.8). It shows a strong, exponential growth below 20K, which has been attributed to a proximity effect coupling neighboring unit cells along the c-axis through poorly conducting and non-superconducting oxide layers (SrO and BiO). A magnetic field can breakdown the coupling if it is larger than a field H_b:

$$H_b = H_b(0) \exp - \frac{d_N}{\xi_N(T)} \tag{9.1}$$

where d_N is the normal layer thickness. $\xi_N(T)$ varies as a power law, and therefore H_b decreases exponentially as the temperature is raised. In this model, the irreversibility field at low temperatures is interpreted as a breakdown field of the coupling between neighboring unit cells along the c-axis.

9.2.2 Applications of 1G wire

1G wire is still at the moment the only HTS wire available in kilometer length, and used in large scale models of power applications. In all of them, it competes against conventional Cu wire. 1G wire is more than one order of magnitude more expensive than Cu wire, but it offers a higher power density that can sometimes justify a substantially higher cost. Costs here are measured in units of $/kA-m. They are of about 10$/ka-m for Cu, and close to 200$/kA-m for 1G wire.

Cables

One of the more advanced applications of HTS wire is for electrical power transport and distribution in urban environment. A typical 1G wire has a width of 4 to 5 mm and a thickness of 0.3 mm (Fig. 9.9). Long length performance is such that it can carry over 100A, which corresponds to more than 200A/cm-width, or an engineering critical current of $1 \cdot 10^4$ A/cm^2. By comparison, the current density in a Cu wire is limited to about $1 \cdot 10^2$ A/cm^2 by the maximum temperature that the surrounding dielectric can sustain without degradation. When the overall size of the cables, including the cryostat for the HTS cable, is taken into account, this advantage of two orders of magnitude is reduced down to a factor of only 3 to 5. This may not sound spectacular, but increasing by such a factor the power that can be transmitted through an underground duct of a given size can be of great interest in urban environments where there is a growing demand for electrical power, and where additional right of way is not easily available.

The cost of the conductor is of about $1 \cdot 10^6$ US$/kA-km, at today's tape price, for a tri-axial cable design where conductors for the three phases are placed in a single cryostat (Fig. 9.9). The cost of the conductor doubles if conductors for each phase are placed in separate cryostats, because in that case each of them must carry its own return current. It is foreseen that in the future the cost of the conductor might come down by a factor of 2 to 4.

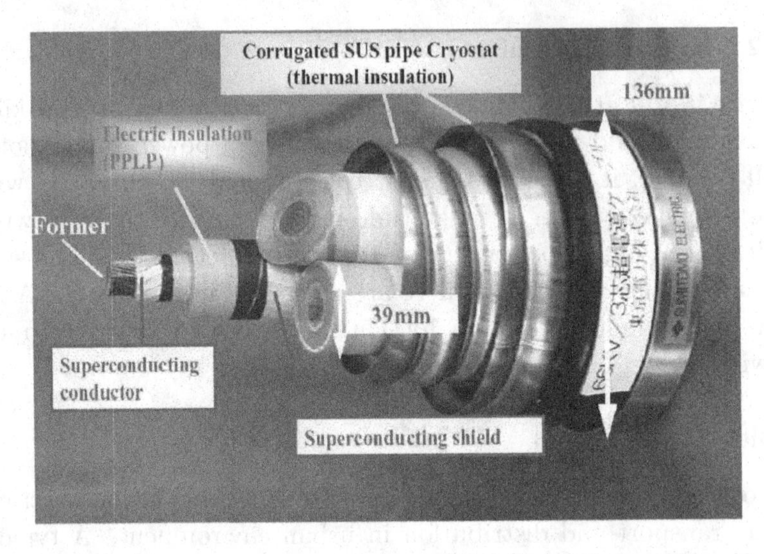

Figure 9.9: Albany team project, IGC-SuperPower, Sumitomo Electric, BOC and Niagara-Mohawk. A Bi-2223 cable with the three phases in a single cryostat. (Courtesy of SuperPower.)

The cost of refrigeration, which includes the costs of the cryostat and that of the cryocoolers, is of the same order as that of the tape. The cost of the cryostat itself is today $0.6{\cdot}10^6$US$/km for a triaxial design, and three times more if each phase is carried in a separate cryostat. It is estimated that this price may come down by a factor of 2 in case of increased production. The cost of the cryocooler depends on the total heat load that must be removed. It includes thermal losses to the environment (3W/m) and a.c. losses within the conductor (1 to 2 W/m). The cost of the cryocooler plus ancillary equipment to remove 1W at 77K is 200 US$, which for total losses of 5W/m translates to $1{\cdot}10^6$ US$/km. Again, that price may come down in the future.

One can expect that the total capital cost of an HTS cable carrying 1000A will therefore be on the order of $1{\cdot}10^6$ US$/km. It does not vary quickly with the power carried by the cable, because the cost of the cryostat and that of the cryocooler do not vary quickly with it (thermal losses to the environment are more or less fixed). It is much larger than that of a conventional cable, except for very high ratings. To decide whether this is a worthwhile investment, one must compare it to that of the additional infrastructure needed to lay down a new conventional cable in an area where more power is needed. That cost will vary considerably with the local environment, since it depends heavily on the cost of land and manpower. In a developed environment, it is quoted in the USA as 5 to $10 \cdot 10^6$ US$/km. This amount will be saved if the new cable can be fitted into an existing duct where it would replace a conventional cable. The increased power density is here the key point, rather than energy savings. Because of the large surface to volume ratio of cables, the cost of refrigeration is a key component of the total cost of an HTS cable. It is already higher than the cost of the HTS conductor, and may become even more dominant as the cost of the latter comes down in the future. Therefore, the global outlook for HTS cables may not change substantially when 2G wire becomes available (see below).

Increasing demands for electricity in fast growing cities is the key driving force behind the development of HTS cable. Demand for such cables will vary considerably from one country to another. By and large, it is higher in the USA and in Asia than it is in Europe. Amongst the current large scale cable projects of which there are about 10 in the world to day, three are located in the USA. To cite one specific example, the Long Island Power Authority will install a 600m long HTS cable, operating at 138kV and 2400A. Three individual envelopes are contained in one single 10" pipe. Total HTS conductor to be used is 128 km.

A different application of HTS cables could be in the future for dc transmission of large blocks of powers over long distances. In that case energy savings can become a primary motivation because the cooling costs of dc HTS cables are basically independent of the transmitted power, the dc losses being essentially zero.

Other applications of 1G wire

1G wire is also being used in demonstrations of HTS transformers, rotating machines and other applications requiring magnets such as research magnets and magnetic separators.

Approximately half of the power lost in transmission and distribution occurs in transformers. Because they are cooled by oil, they are prone to fire hazards and additional space is needed for protection. Occasional overload reduces the life-time of the dielectric. HTS transformers offer the advantages of presenting no fire hazard since cooling is cryogenic, no life-time shortening because of overload, increased efficiency, smaller weight and reduced space needed. A 5 MVA HTS transformer was built and operated successfully by a team at Wankesha, USA. Its coils were wound with 1G wire and operated at 30K. It reduces voltage from 24.9 KV down to 4.2 KV. It was concluded that scale up to 30MVA (138 KV down to 13.8 KV) will require an improved (lower ac losses, higher j_c) and cheaper conductor than 1G wire, as well as better cold dielectrics.

In generators, dc magnetic fields generated by the rotor coils cut through the windings of the stator, to produce ac voltage and power. dc currents in the rotor windings and ac currents in the stator windings are also a characteristic of electrical motors. Because dc losses in superconductors are practically zero below the critical current, while ac losses are not because vortices can "vibrate" around pinning defects, superconducting windings in generators and motors are used in the rotor rather than in the stator, even though cooling the rotor windings is obviously more complicated than cooling the stator ones. Benefits of using superconducting windings include reduced losses, elimination of iron cores used in conventional rotating machinery to enhance the magnetic field, smaller size and weight. For generators, even the modest 0.5% improvement in efficiency can justify the increased capital cost over the life time of the machine. In HTS motors, savings might even pay for the entire capital cost over its life time.

The main technical requirements from the HTS windings in rotating machines are that they must generate fields of a few Tesla, and in the case of large scale generators to be able to withstand the centripetal forces on the rotor. 1G wires can generate the desired fields if they are cooled down below 30K. They can also withstand the mechanical stresses in slowly rotating machines such as large scale

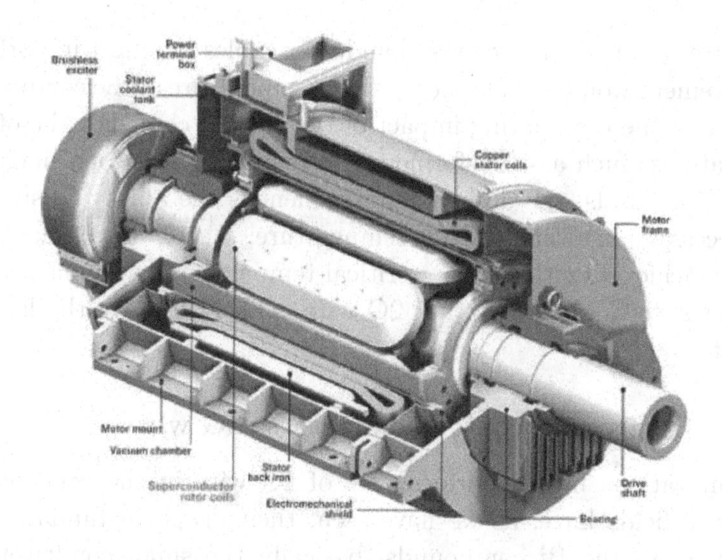

Figure 9.10: Large scale superconducting motors (several MW power) have been demonstrated with Bi-2223 1G wire. The rotor is superconducting and the stator normal. (Courtesy of American Superconductor.)

motors, but not in large scale generators where the windings must sustain 10,000 g on a 1 meter diameter rotor. Several 1000 hp motors have been constructed and operated successfully with 1G wire. This corresponds to the typical rating of large scale motors used in heavy industries, which consume up to 20% of its total electricity needs. However, from an economic standpoint, the cost of 1G wire and that of refrigeration are higher than the value of the energy savings over the life time of the machine. There remains the advantage of smaller size and weight, which can be important when space and weight are at a premium, as for the case on board ships and aircraft (Fig. 9.10). Prospects for a larger use of HTS motors and generators will improve with the availability of 2G wire, see below.

The above brief review of the possible applications of 1G wire have brought to the front a basic distinction between cables on the one hand, and applications where the conductor must sustain a field in the Tesla range, on the other hand. For cables the most important improvement would be a higher critical temperature of the conductor, because of the dominating impact of the cooling cost. For the other applications, such as transformers, motors and generators, magnets of different kinds, the desired improvement is the ability to sustain the necessary field at a higher temperature, a goal that is not necessarily achieved with a higher critical temperature as we have seen. It is for such applications that 2G wire is likely to have the largest impact.

9.2.3 Progress in coated conductors: 2G wire

The limitations in the performance of 1G wire in the presence of magnetic fields have, as we have seen, their origin in fundamental properties of the Bi compounds, basically the small condensation energy per coherence volume. This has led to considerable efforts towards the development of conductors based on YBCO, which does not suffer from the same limitation. Long length 2G wire is not yet commercially available, and contrary to the case of 1G wire for which manufacturing processes are now well established, a number of different processes for coating YBCO films on metallic substrates are still under active consideration.

Methods of deposition

Methods of YBCO film deposition include: physical vacuum deposition methods onto heated substrates (sputtering or pulsed laser deposition (PLD) from alloy targets, simultaneous deposition of the constituents); chemical methods including Metal Organic Chemical Vacuum Deposition (MOCVD) where organo-metal molecules of the constituents are deposited onto the heated substrate where they decompose while YBCO forms at the same time, and Metal Organic Deposition (MOD) where the molecules are deposited at room temperature, the formation of YBCO being performed at a later stage. (See A. Goyal in suggested further reading).

All of these methods can lend high quality films having j_c values of more than $1{\cdot}10^6$ A/cm^2 at 77K, when films are grown on single crystal substrates such as $SrTiO_3$ having a good lattice match and that do not react with YBCO. Selection of the method of deposition to be used in an industrial process will therefore be based mostly on economic considerations (which includes of course fabrication yield). Because of the nature of the substrates (described below) that must be used for the production of long length tapes, additional "seed" or "buffer" layers must be grown on them before the YBCO layer itself. These layers are all oxides of different kinds (Y_2O_3, Yttrium Stabilized Zirconia or YSZ, CeO_2). Their growth methods are similar to those used for the YBCO layer itself. All methods of tape fabrication are reel to reel processes.

PLD and sputtering methods have the advantage that they use alloy targets whose exact composition is preserved in the deposition process. There is in principle no thickness limitation. They lend themselves to variants such as successive in-situ depositions (multilayers) and controlled introduction of additional oxides for the formation of nano-dots and other pinning defects. Their main drawback is the high cost of vacuum equipment and the relatively slow growth rate of the deposited layers, resulting in a high capital cost per unit length of coated conductor. Vacuum co-evaporation is an alternative that provides faster growth rates.

MOCVD is an in-situ deposition and growth process that combines some of the advantages of physical vacuum deposition (no thickness limitations, possibility of multi-layers) with a lower capital cost (cheaper equipment, faster deposition rate). Linear tape production speeds of several 10 m/hr have been demonstrated. By running the tape as an helix in the deposition zone, the desired tape width can be produced directly without further slitting, while keeping up the high throughput (Fig. 9.11).

The principal advantage of the MOD process is that it uses much cheaper precursors than MOCVD. The reaction leading to YBCO formation is:

$$2BaF_2 + 2H_2O + 0.5Y_2O_3 \rightleftarrows YBCO + 4HF$$

MOD combines lower equipment cost and lower materials cost, compared to physical vacuum deposition and MOCVD. One of its

Figure 9.11: Helix tape handling in an MOCVD deposition chamber allows the direct deposition on the required tape width without losing deposition speed. (Courtesy of SuperPower.)

disadvantages is that it is a single step process. HF molecules are released during annealing and must be removed during YBCO formation, which limits the thickness of films that can be produced, since removal takes place through diffusion. Because annealing is a relatively slow process, high throughput must be achieved by deposition on broad substrates, to be slit at a later stage. Current plans contemplate deposition on a 10 cm width tape, to be slit into 4 mm conductor strips.

Substrates and buffer layers

Two different kinds of substrates have been successfully developed to achieve the required high degree of grain orientation: highly textured rolled cubic NiW alloy tapes, the so-called "RABiTS" (Rolling Assisted Bi-axially Textured Substrate) substrates; and polycrystalline Haste alloy tape coated with a MgO thin film whose orientation is achieved by an ion assist beam during deposition, the so-called Ion Beam Assisted Deposition or IBAD. Lengths of several 100 meters of well oriented substrate tapes of both kinds have been produced successfully (even a few kilometers in the case of RABiTS).

RABiTS substrates: Tungsten is added to Nickel to improve mechanical properties and reduce magnetism. Typical W concentrations used are in the range of 3 to 5%. Attempts at eliminating magnetism altogether by increasing the W concentration beyond 5% have been unsuccessful because the cubic structure is not preserved. X-Ray Phi scans have shown a Full Width at Half Maximum (FWHM) corresponding to less than 5° in-plane misorientation between NiW grains.

The stack of buffer layers consists of a thin layer of Y_2O_3 (30 nm) to prevent transformation of Ni into NiO by oxygen diffusing out from the next YSZ thicker layer (200nm) which provides a suitable lattice for hetero-epitaxial growth of YBCO, and an additional thin layer of CeO_2 (30 nm) that gives a smooth surface finish with a surface roughness of less than 1 nm. Orientation of the NiW tape is preserved through the stack of buffer layers, with the FWHM measured on the CeO_2 corresponding to a misorientation of 5° to 6°. 100 meters long, 4 centimeters wide buffered NiW tapes are now routinely produced.

The FWHM gives only an averaged misorientation. A more detailed and interesting information is that of the individual GB misorientation since it controls the local critical current. Such misorientation maps have been established. For the CeO_2 cap layer they show that 70 to 80% of the GBs have a misorientation of less than 4°. For the YBCO layer, special attention has been given to the in-plane component of the misorientation. Again using the 4° threshold, it has been shown that in 70% of the film GBs do not limit the critical current. A 0.12 μm thick film of such quality has a critical current density of $4{\cdot}10^6$ A/cm^2 at 77K.

IBAD substrates: In the IBAD process, the stack of buffer layers consists typically of the following: a polycrystalline layer deposited on the Haste alloy substrate acting as a diffusion barrier with the rest of the stack (80 nm Al_2O_3); a Y_2O_3 nucleation layer (7 nm); the IBAD layer (10 nm YSZ or more recently MgO; homo-epitaxial MgO

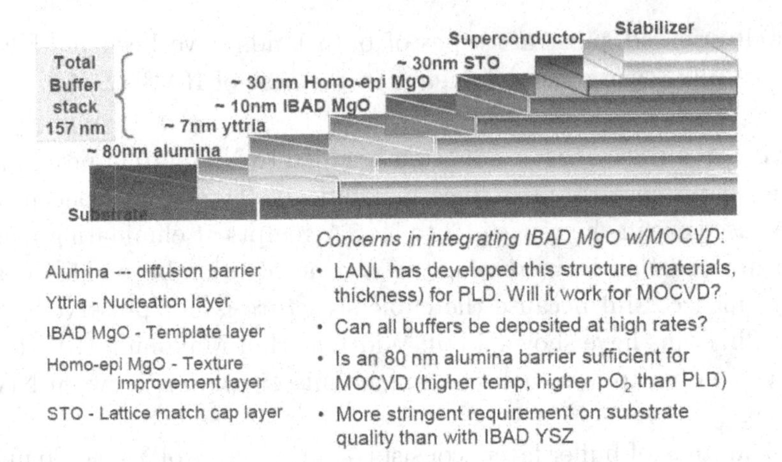

Figure 9.12: A typical IBAD stack used in the MOCVD process. (Courtesy of Superpower.)

for texture improvement; a $SrTiO_3$ layer for lattice match improvement (30 nm)) (Fig. 9.12). All layers of the stack can be prepared in-situ in the MOCVD process. Growth temperatures are higher than in other growth methods (such as PLD), therefore the role of the alumina layer to prevent diffusion is essential. Quoted misorientations are similar to those obtained with RABiTS substrates.

9.2.4 Coated conductor performance

YBCO films grown on well lattice matched single crystal substrates such as $SrTiO_3$ or $LaAlO_3$ achieve j_c values of several $1{\cdot}10^6$ A/cm^2 (up to $5{\cdot}10^6$ A/cm^2) at 77K, and several $1{\cdot}10^7$ A/cm^2 at 4.2 K for film thickness of a few 100nm. At 77K, j_c drops down by about one order of magnitude under a field of 3T, and two orders of magnitude at 6T. This is a huge improvement over 1G wire. In this section, we first specify the performance goals expected from 2G wire, and review the main results obtained so far in self-field and under applied fields.

Performance goals

From an engineering standpoint, tape performance is measured in A/cm-width. Performance goals have been defined for each application as follows:

(i) for cables, the goal for broad commercial application is 500 A/cm-width at 77K. This is to be compared to 130 A/cm-width, the best performance at the time of writing of 1G wire over long length (on the order of 100m).
(ii) for motors and generators, the goal ranges from 100 A/cm-width at 65K to 450 A/cm-width under 3T at 30K, depending on the specific application.
(iii) for high field coils operating at liquid Helium temperature (research magnets, NMR magnets, plasma confinement), the goal is to improve the highest field reached or planned today (more than 20T for NMR magnets, 12T for the large coils of the ITER project). HTS coils could be in the form of inserts placed inside LTS coils.

Self-field performance — thickness dependence

The goal of 500 A/cm-width for cable applications corresponds to j_c=5·10^6 A/cm^2 for a 1μm thick film. The problem that needs to be overcome in order to reach that goal is to decrease the critical current density with film thickness (Fig. 9.13).

In high quality, well oriented films in which GBs play a negligible role, such as are now available as described above, $j_c(t)$ has been found to follow a universal behavior, independent of the method of deposition. It extrapolates to a value of 7·10^6 A/cm^2 at zero thickness, and saturates at about 1.5·10^6 A/cm^2 at large thickness $t > 3\mu$m. The measured $j_c(t)$ has been modeled by a linear incremental dependence where it varies between these two values between zero thickness and 0.65μm, and stays constant beyond that. To achieve the desired goal would then require a thickness of 3 to 4 μm, meaning three times more material and three times longer deposition time, in the end an almost three times higher production cost compared to what it would be for a 1 μm thick film. For the MOD process, it

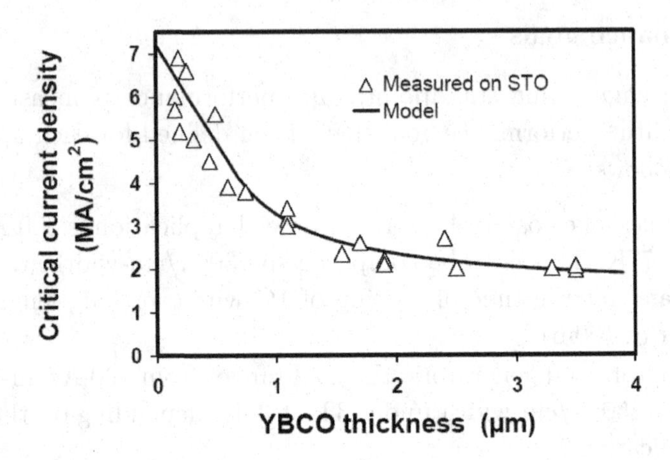

Figure 9.13: Decrease of the critical current density with YBCO PLD deposited film thickness. (Courtesy of S. Foltyn, Los Alamos National Laboratory.)

may in fact not be possible at all to reach such large thickness and still have a high quality film.

One should first note that the zero thickness extrapolated j_c is close to the depairing limit. In Sec. 8.7.1 we calculated a depairing limit of $1{\cdot}10^7$ A/cm^2 for YBCO at 77K. Therefore no significant improvement in the achieved $j_c\,(t \longrightarrow 0)$ is to be expected. It is the strong thickness dependence that needs to be understood and overcome.

Even in self-field, dissipation is due to the motion of vortices. One interpretation for the thickness dependence is that j_c values near the interface are enhanced by pinning from extended defects, while smaller and less effective defects in the bulk are responsible for the smaller value at large thickness. Extended defects near the interface could be, for instance, misfit dislocations generated by the difference between the lattice parameter of YBCO and that of the last buffer layer. The crucial role of extended defects to pin vortices in the HTS has been discussed in Sec. 8.8.3. They are needed to compensate at least in part for the relatively low condensation energy per coherence volume.

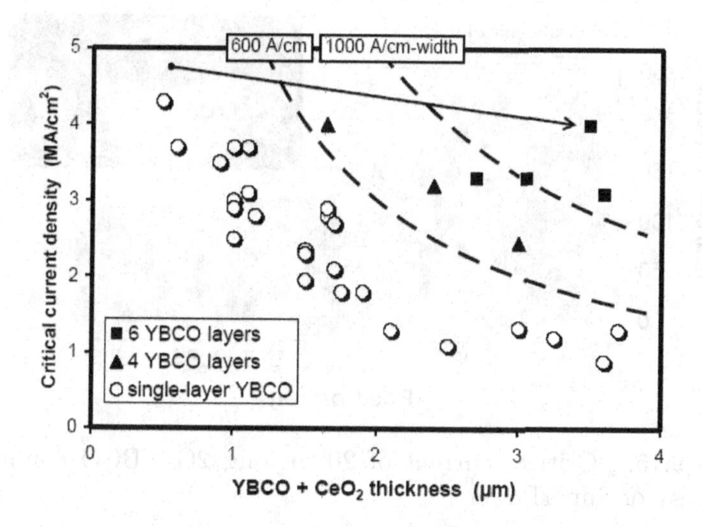

Figure 9.14: The critical current density can be considerably increased by using multi-layers which multiply interfaces. (Courtesy of S. Folstyn, Los Alamos National Laboratory.)

If this interpretation is correct, a way to increase j_c in thick films would be to introduce additional interfaces. This is what has been done by a team at LANL (Fig. 9.14). Thin CeO_2 layers were intercalated in YBCO during growth. An average j_c=$4{\cdot}10^6$ A/cm^2 has been obtained in a 2 μ m thick film. Such a film has a critical current of 1000 A/cm-width on short samples, which is a realistic value towards the desired goal of 500 A/cm-width over long lengths. This intercalation process was performed by PLD, but is compatible with the more industrial oriented MOCVD process, although may be not with MOD. Other ways to introduce additional interfaces do exist, such as nano-scale precipitates of non-superconducting oxides (for instance $BaZrO_3$). The remarkable achievement of coated conductors research and development programs is that after improving the grains' alignment to the point where the detrimental effect of grain boundaries has been eliminated, one has now reached the stage where the critical current of the coated conductor is being improved beyond

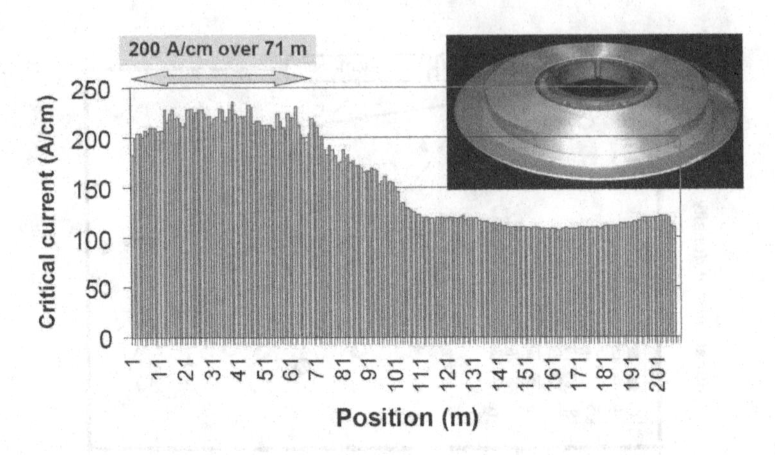

Figure 9.15: Critical current of 200m long 2G YBCO conductor. (Courtesy of SuperPower.)

that of the intra-grain value in good quality films by the controlled introduction of extended defects. As we shall see below, this has also been a decisive step towards the application of 2G wire under applied fields.

At the time of writing, current long length (100m) performance of YBCO coated conductor is typically 100 A/cm-width at 77K, still 5 times smaller than the desired goal. 200 A/cm-width over 70 meters has been achieved by a team at Superpower (Fig. 9.15). The set goal appears to be within reach with multi-layers or other methods to introduce extended defects.

Performance at intermediate fields

At intermediate field values, up to a field $H_{cr}(T)$, the critical current has been found to follow a power law of the form $j_c \propto H^{-\alpha}$. In as-grown films, $\alpha \propto 0.5$ is almost independent of temperature. The goal of 450 A/cm-width at 3T and temperatures not lower than 30K is not reached either in these films at 40K (both at about 220 A/ cm-width) (Fig. 9.16). Single layer YBCO 2G wire does have an enormous advantage over 1G at higher temperatures and moderate

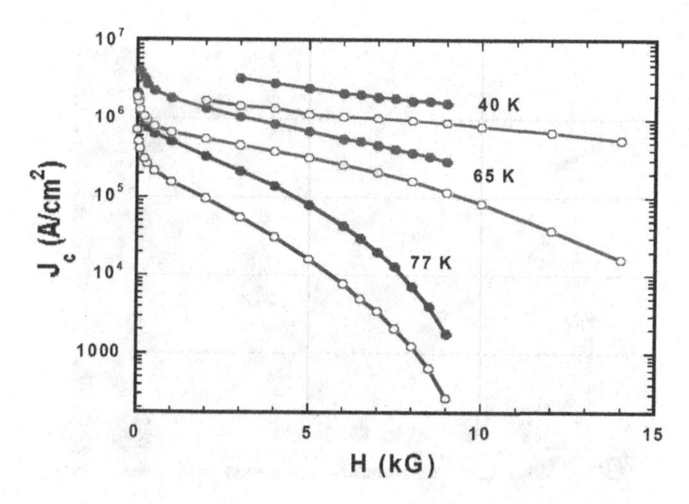

Figure 9.16: Empty symbols, as grown films. Full symbols, nano-rods doped films. (Courtesy of A. Goyal, Oak Ridge National Laboratory.)

fields, for instance at 65K and under 1T, 2G wire can carry 200 A/cm-width, and 1G wire only a few A/cm-width. But this advantage is not of great interest for most coil applications for which a field of 1T is insufficient.

Fortunately, the performance of 2G coated conductor (still over short lengths) has recently been considerably improved by the inclusion of $BaZrO_3$ nano-dots and other extended defects. These nano-dots have been found to self-align in the direction perpendicular to the CuO_2 planes forming nano-rods of a few nanometers diameter, regularly spaced in the YBCO matrix (Fig. 9.17).

One remarkable effect of these nano-rods is to reduce considerably the field anisotropy of $j_c(H)$. In as-grown films j_c presents a peak in the field orientation parallel to the CuO_2 planes and a minimum in the perpendicular one, the anisotropy ratio being of about 5. In a coil configuration, it is the lower value that will control the critical current since all field orientations will in fact be present. In films containing controlled extended defects such as self-aligned nano-dots, this anisotropy is eliminated, and in some cases it is even reversed.

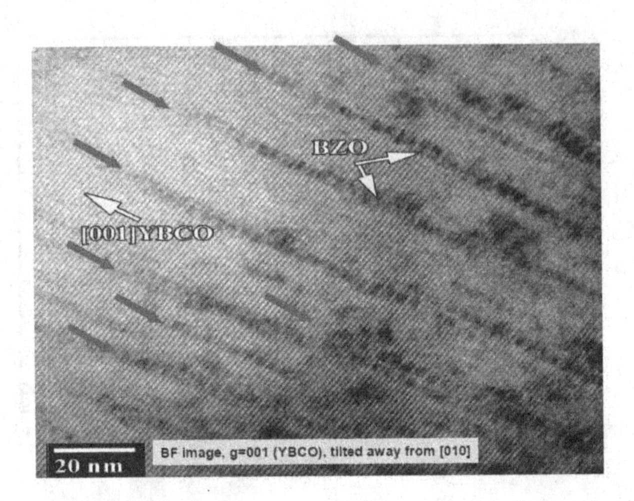

Figure 9.17: Doping YBCO films with a few % of $BaZrO_3$ produces nano-rods that can pin vortices very effectively. (Courtesy of A. Goyal, Oak Ridge National Laboratory.)

At the same time, the exponent α is reduced from 0.5 in the as grown film, to about 0.3, showing that in as-grown films it is the lack of vortex pinning in the perpendicular orientation that is responsible for the insufficient performance (Fig. 9.18).

With the incorporation of nano-rods, the field dependence of the critical current becomes extremely weak. This is a remarkable illustration of the improvement that can be reached with the help of extended defects parallel to the applied field, as discussed in the preceding chapter. For instance, in a BZO doped 1μ m thick film, at 65K j_c decreases only from $4{\cdot}10^6$ A/cm^2 in self-field down to $2{\cdot}10^6$ A/cm^2 at 1 T At 40K, such a film meets easily the required performance of more than 450 A/cm-width under 3T. Multi-layering also reduces the anisotropy and leads to the required performance.

In the end, the fundamental difference between 1G and 2G wires is that controlled extended defects can improve the performance of the latter but not that of the former, at the temperatures relevant for motor and generator applications. This is both an empirical conclusion (defects do not enhance the value of $H_{\rm irr}$ in the Bi compounds in

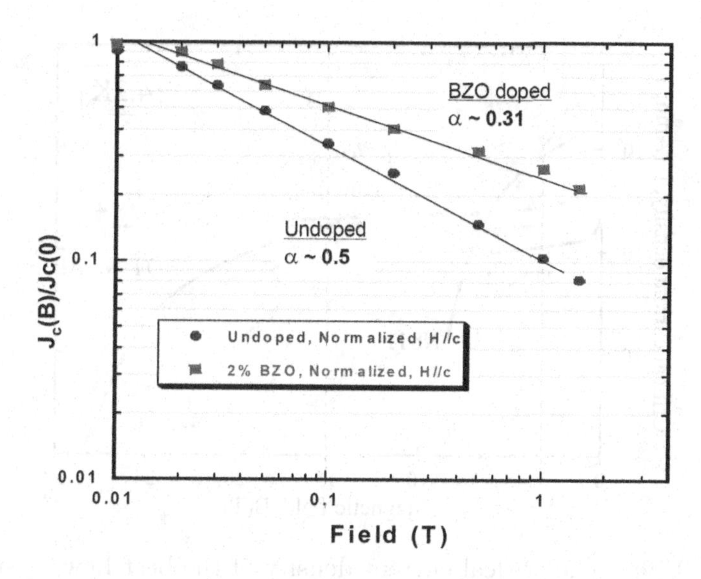

Figure 9.18: The field dependence of the critical current is considerably reduced by the presence of nano-rods. (Courtesy of A. Goyal, Oak Ridge National Laboratory.)

that temperature range, while they do enhance it in YBCO), and one with a solid theoretical foundation: the fact that the condensation energy per coherence volume in the Bi compounds is just too small compared to $k_B T$, even at reduced temperatures of the order of 0.3, to allow extended defects to improve $j_c(H)$. It is interesting to compare the theoretical criterion for loss of vortex rigidity defining the limiting field H_L, and the highest value of $H_{\rm irr}$ achieved in coated conductors. In Sec. 8.4.2 we calculated H_L=0.8H_{c2} at 77K. For BZO doped YBCO, taking into account the reduced value of T_c=88K, we estimate $H_{c2}(T$=77K)=15T, which gives H_L=12T, while the highest value of $H_{\rm irr}$ reported for YBCO films with nano-rods is close to 10T (which is similar to values reported by Figueras *et al.*).

Low temperature — high field performance

At low temperatures, limitations of the critical current due to thermal fluctuation become unimportant. We can then expect critical

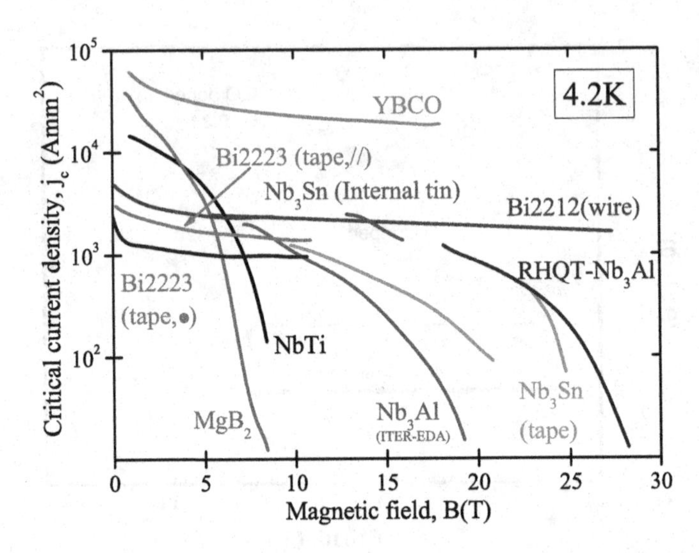

Figure 9.19: The critical current density of the best Low T_c superconductors falls off below 25T, while that of the High T_c stills changes very little with field, opening up the prospect of higher field magnets.

current densities to remain high up to fields of the order of $H_{c2}(0)$, which as we have seen in the previous chapter is for the cuprates of the order of 100 T. This is five times higher than in the best HTS, and opens up the possibility of a new generation of superconducting magnets.

Figure 9.19 shows the field dependence of the critical current density at 4.2K in LTS wires or tapes used today in magnets, compared to the performance of HTS superconductors. For NbTi it collapses near 10 T, for Nb_3Sn near 25 T. By contrast, it remains almost constant beyond 20T in Bi 2212 and YBCO. In fact, we do not know at the present time what are the limits of HTS superconductors at 4.2 K, simply because there are not powerful enough magnets to perform critical current measurements beyond 30T. What is remarkable is that in YBCO for instance critical currents remain at the level of more than $1{\cdot}10^6\,\mathrm{A/cm^2}$ up to more than 20T. This critical current density is two orders of magnitude higher than in Nb_3Sn at that field. Substantial improvement in the highest fields provided by supercon-

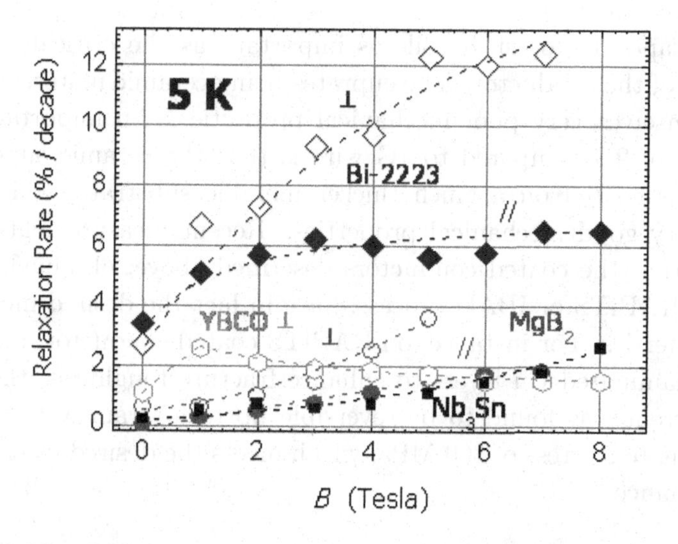

Figure 9.20: Magnetization relaxation rates for a variety of low and high T_c superconductors. Note that at high fields YBCO has the lowest relaxation rate. After C. Senatore, P. Lezza and R. Flükiger, to be published.

ducting magnets can be expected once HTS tapes are available as long length conductors.

In most high field applications, superconducting coils are used in the persistent mode where the coil is closed upon itself (NMR, magnets for high energy physics research, plasma confinement). As discussed in Sec. 6.4, flux creep must be very slow to allow such applications. Very recently, flux creep data obtained from magnetization relaxation measurements have shown that 2G wire is far superior to 1G wire at any field, and to low temperature superconductors at very high fields (Fig. 9.20). The data shows that the flux hopping rate is much smaller in YBCO than in the bismuthates, consistent with the difference of about one order of magnitude between the condensation energies in these two superconductors as discussed in Sec. 3.2.

For large scale coils operating at high fields, the magnetic pressure imposes considerable stress on the material. The resulting strain can produce irreversible damage to the coil. Mechanical properties of the tape are for such coils as important as the critical current density of the conductor. The cuprates being ceramic materials have by themselves very poor mechanical properties. An important advantage of 2G compared to 1G wire is that the ceramic superconductor is coated on a much thicker metallic substrate, which can have very good mechanical properties. Recent measurements have shown that the coated conductors described above, obtained either by the RABiTS or IBAD process, have in fact the desired mechanical properties. For instance, on RABiTS coated conductor, incorporating laminated Cu layers to enhance fracture toughness, the critical current was found to be reversible up to a strain of 0.4 corresponding to a stress of 300 MPa, which meets the desired mechanical performance.

9.3 Further reading

For a review of grain boundary critical currents see H. Hilgenkamp and J. Mannhart, Rev. Mod. Phys. **74**, 485 (2002).

For a description of the various methods of fabrication of 2G HTS conductors see "Second-Generations HTS Conductors", ed. Amit Goyal, Kluwer Academic Publishers (2005).

Index